MEXICO IN SPACE

MEXICO IN SPACE

FROM LA RAZA CÓSMICA TO THE SPACE RACE

ANNE W. JOHNSON

THE UNIVERSITY OF
ARIZONA PRESS
TUCSON

The University of Arizona Press
www.uapress.arizona.edu

We respectfully acknowledge the University of Arizona is on the land and territories of Indigenous peoples. Today, Arizona is home to twenty-two federally recognized tribes, with Tucson being home to the O'odham and the Yaqui. The university strives to build sustainable relationships with sovereign Native Nations and Indigenous communities through education offerings, partnerships, and community service.

ISBN-13: 978-0-8165-5484-3 (hardcover)
ISBN-13: 978-0-8165-5483-6 (paperback)
ISBN-13: 978-0-8165-5485-0 (ebook)

Cover design by Leigh McDonald
Cover art: "De la Diana a la Luna/From the Diana to the Moon." Artist: Ana Laura Contreras Madrigal, winner in the open category of the Third Space Art Contest "Mexico to the Moon" (2018) sponsored by the AEM. Courtesy of Mario M. Arreola-Santander, Outreach Director of the Mexican Space Agency.
Interior designed and typeset by Sara Thaxton in 10.5/14 Warnock Pro with Helvetica Neue LT Std and Korolev Condensed

Library of Congress Cataloging-in-Publication Data
Names: Johnson, Anne W., (Anne Warren), 1971– author
Title: Mexico in space : from la raza cósmica to the space race / Anne W. Johnson.
Description: [Tucson] : University of Arizona Press, 2026. | Includes bibliographical
    references and index.
Identifiers: LCCN 2025029991 (print) | LCCN 2025029992 (ebook) | ISBN 9780816554843
    hardcover | ISBN 9780816554836 paperback | ISBN 9780816554850 ebook
Subjects: LCSH: Astronautics—Mexico | Astronomy—Mexico
Classification: LCC TL789.8.M6 J64 2026 (print) | LCC TL789.8.M6 (ebook)
LC record available at https://lccn.loc.gov/2025029991
LC ebook record available at https://lccn.loc.gov/2025029992

Printed in the United States of America
♾ This paper meets the requirements of ANSI/NISO Z39.48-1992 (Permanence of Paper).

CONTENTS

List of Illustrations   *vii*
Acknowledgments   *ix*
Abbreviations   *xiii*

Introduction   3

1. Cosmohistories   25
   Interlude: *Nepantla* Space Program   65
2. Mexico in Orbit   67
   Interlude: *La NASA no es la raza*   97
3. The Space Generation   99
   Interlude: *Matters of Gravity*   129
4. In the Navel of the Moon   133
   Interlude: The Great North American Eclipse   169
5. Transhabiting Mars   173
   Interlude: Mars Station   205
6. Dark Skies   207

Notes   *239*
References   *251*
Index   *269*

ILLUSTRATIONS

1. Launching rockets   7
2. Sargasso   15
3. *Duality*   30
4. Piedra del Sol   40
5. Coatlicue   46
6. *Autonomous InterGalactic Space Program*   66
7. *México, ¡me subo a tu nave!*   69
8. Tulancingo 1   95
9. *La NASA no es la raza*   98
10. Genaro Grajeda, MEX-1 Satellite Officer   118
11. *Supernova*   131
12. *Los sueños se construyen sobre las raíces*   134
13. Colmena   150
14. Apollo astronauts in El Pinacate   156
15. Eclipse   171
16. Martenochtitlan and Martelolco   178
17. *Codex Martenochtitlan*   189
18. Mars Station   206
19. Light pollution   214
20. National Observatory, Tonantzintla   219
21. Peña del Aire   225

ACKNOWLEDGMENTS

I have depended on so many people throughout the process of research-ing and writing this book that it's hard to know where to begin. Following the logic of the milieu, I suppose the best way to organize the expres-sions of gratitude is with clusters on an imagined rhizome of relations. At the AEM, I owe a great debt to the knowledge and generosity of Carlos Duarte and Mario Arreola, who were among the first people to show interest in this project. I have also benefitted immensely from conver-sations with Adán Salazar, Alberto Lepe, Kesniel Bravo, and Amanda Gómez at the AEM. Representatives of the AFAC and FEMIA were also very helpful as I sought to understand Mexican technological, political, and economic aspirations.

Nahum, Manuel Díaz, and Mariana Paredes at the KOSMICA Institute have been incredible interlocutors, as have artists Ale de la Puente, Mar-cela Armas, Juan José Díaz Infante, Ilana Boltvinik, and Rodrigo Viñas. Marcela Chao and her fellow interplanetary travelers at Marsarchive .org, particularly Juan Claudio Toledo, Amadís Ross, Alejandra Espino, and Raúl Santos, injected a serious playfulness into my research, without which this project would have been radically different. Johannes Neur-ath gave me great insight into pre-Hispanic and Indigenous perspectives on the cosmos. And discussions with artist Raúl Cruz Figueroa, whose paintings fuse fantastic versions of the past and the future, have helped me think about slippery times and identities from Mexico.

Oscar Ojeda of the Cydonia Foundation in Bogotá has been a long-time collaborator and enthusiastic reader of my work. Current and former SGAC members, like Tania Robles, Kaori Becerril, Emilie Karina Estrada, Itzel Rocío, Juan Carlos Mariscal, and César Augusto Serrano, have also been amazing sources of generational perspectives often missing in other spaces. At the UNAM-LINX, Gustavo Medina Tanco and his students have been fonts of information and analysis that have been fundamental for this project. José Franco and Antígona Segura, also of the UNAM, were early interlocutors who helped me contextualize the space sciences in Mexico. And Bernard Foing of the IAF (among other organizations) has been a font of knowledge and alternative perspectives on human futures in space.

The work of Mariana Domínguez and Juan Antonio Moreno of MAR D' SAL inspired one of this book's central themes, for which I am truly grateful. Ramón Córdova and Felipe Ávila, the organizers of the ENMICE rocketry competition, were unexpected collaborators who provided me with access to yet another dimension of Mexican space activities.

As I became more and more fascinated with the right to dark skies in Mexico, Joshua Muñoz became a friend and collaborator. And the *socios* of Peña del Aire have been incredible teachers, as well. Crescencio, Angélica, Ana, Silverio, Silvestre, Noé, Marcelo, and many others became, in many ways, the heart of this project, and I hope to continue working with their community for years to come.

Members of the Social Studies of Outer Space network (SSOS) and other working groups and research projects, both ephemeral and long-term, have provided a community of like-minded scholars to whom I've never had to justify doing what I do. Particular thanks for fruitful exchanges are due to Eleanor Armstrong, Fabiane Borges, Marie-Pier Boucher, Victor Buchli, Carmen Bueno, Giles Bunch, Marta Cabrera, Santiago Carassale, Paola Castaño, Felipe Cervera, Javier Contreras, Oliver Dunnett, Vanessa Freije, Réka Gál, Alexander Geppert, Maritza Gómez, Jenia Gorbanenko, Alice Gorman, Oscar Guarín, David Jeevendrampillai, Karlijn Korpershoek, Adryon Kozel, Willi Lempert, Cristina Luna, Lilián Martínez, Javier Mejuto, Lisa Messeri, Carlos Mondragón, Evan Moritz, Jorge Mujica, Andrea Murillo, Hanna Neiber, Chakad Ojani, Valerie Olson, Brian Odom, Daniela Ortega, Aaron Parkhurst, Joseph Popper, Lauren Reid, Delázkar Rizo, Juan Francisco Salazar, Javier Serrano, Anna

Szolucha, Brad Tabas, Peter Timko, David Valentine, Zinaida Vasilyeva, Alejandro Viadas, Matjas Vidmar, Daniel Vizuete, and Nina Witjes.

At my own institution, the Universidad Iberoamericana in Mexico City, I must mention the support and collaboration of my colleagues Lorena Arias and Gerardo Martínez of the Clavius Observatory, as well as the other members of the Ibero's incipient Space Humanities seminar: Edward Bermúdez, José María Nava, María Zorilla, Pilar Álvarez, Eduardo Vega, and Ale Quintana. Between 2020 and 2022, and again from 2025 to 2027, the "Ibero" provided funding for part of the project through its directorate of research. I am also very grateful to José Luis Barrios and the CEX for sponsoring the Marsarchive.org exhibition I describe in chapter 5. My colleagues in the anthropology department have my undying gratitude for accepting my weird obsessions with grace and curiosity. My friends Rodrigo Díaz and Melissa Biggs provided useful feedback at different stages of the book's writing and continue to endure hours of conversation about its themes. As this book passed from proposal to final manuscript, its anonymous reviewers not only made extraordinarily valuable suggestions for deepening and clarifying arguments but also served as cheerleaders for the project. Editor Allyson Carter was an enthusiastic supporter from the beginning. She and her editorial assistant Alana Enriquez at University of Arizona Press were thoughtful stewards of the project throughout the editorial process. This book also benefitted enormously from Heather Jacobson's amazing copyediting services.

The discussion of Mexican satellites that appears in chapter 2 appears in a much shorter version in the *Routledge Handbook of Social Studies of Outer Space* (2023), and an abbreviated description of the activities of Marsarchive.org that I include in chapter 5 was published in *Outer Space and Popular Culture* (2023).

Finally, my daughters Catherine and Clara de la Puente have inspired and supported me, putting up with extensive research trips and long writing hours. This book is, as always, dedicated to them.

ABBREVIATIONS

ACT	Art, Science, and Technology (UNAM)
AEM	Mexican Space Agency
AFAC	Federal Civil Aviation Agency
ALCE	Latin American and Caribbean Space Agency
ATDT	Agency for Digital Transformation and Telecommunications
CDC	Center for Digital Culture
CENALTEC	Training Center in High Technology
CFE	Federal Commission of Electricity
CONACYT	National Council for Science and Technology
CONEE	National Council for Outer Space
COPUOS	Committee on the Peaceful Uses of Outer Space (UNOOSA)
ENMICE	Mexican Encounter of Experimental Rocketry Engineering
FEMIA	Mexican Federation of the Aerospace Industry
IAC	International Astronautical Congress
IAU	International Astronomical Union
IAF	International Astronautical Federation
IFT	Federal Institute of Telecommunications
IGY	International Geophysical Year
INAOE	Institute of Astronomy, Optics, and Electronics

INMUJERES	National Institute of Women
IPN	National Polytechnic Institute
ISRO	Indian Space Research Organization
ISU	International Space University
ITAR	International Traffic in Arms Regulations
LANAE	National Laboratory for Access to Outer Space (UNAM)
LINX	Laboratory of Space Instrumentation (UNAM)
MEXSAT	Mexican Satellites
MDRS	Mars Desert Research Station
NASA	National Aeronautics and Space Administration
OAS	Organization of American States
PEM	Mexican Space Program
PEU	University Space Program (UNAM)
PEMEX	Mexican Oil
SCT	Secretary of Communication and Transportation
SECTUR	Secretary of Tourism
SEDENA	Secretary of the National Defense
SGAC	Space Generation Advisory Council
SRE	Secretary of International Relations
UASLP	Autonomous University of San Luis Potosí
UNAM	National Autonomous University of Mexico
UPAEP	Popular Autonomous University of the State of Puebla

MEXICO IN SPACE

INTRODUCTION

All the people sent by the multibillionaire Elon Musk to inhabit the city were dead and dried up.

—MARTENOCHTITLAN, FIRST FOUNDING

Space Is for Everyone?

This slogan is everywhere in the global space industry, suggesting that "outer space," particularly in the commercial, "democratic" discourse that characterizes the NewSpace economy, is immense and expansive enough for "everyone" to participate in its exploration and exploitation. While meant to be inclusive and inspirational, the phrase implies a world in which there is one, singular outer space available for appropriation by a singular humanity. For the past eight years, I have explored the ways in which a variety of social actors in Mexico—a country not usually associated with space-faring activities—have engaged with outer space in ways that challenge the spatial and agentive singularities implied in both "space" and "everyone."

As Lesley Byrd Simpson (1966) memorably argued in a work originally published during World War II, there are "many Mexicos." In this book, I show how these multiple Mexicos have generated a multiplicity of "outer spaces" capable of fragmenting the totalizing effects of "space is for everyone." In the following chapters, I use the concept of *space milieux* to tell stories about historical and contemporary Indigenous conceptualizations of the cosmos, artistic explorations of science and technology, governmental space agencies promoting national development, cultural collectives that speculate about possible futures on Mars, university laboratories designing lunar robots, associations and start-ups meant to provide young people with a future in the space industry, and communities attempting to defend their territories and provide economic alternatives to immigration through the promotion of astrotourism and the protection of dark skies.

Some of the narratives generated by these groups faithfully repeat Western narratives about space exploration while others are explicitly confrontational, but most interact with hegemonic discourses in complex and contradictory ways. In these stories, for example, the promises of a "Mexican Silicon Valley" and the potential of a European-style Latin American space agency co-exist with issues of Indigenous rights and environmental justice, as well as powerful conceptualizations of the relations between terrestrial and cosmic phenomena that do not fit neatly into Western scientific frameworks. Taken together, the chapters are meant to offer suggestions for how outer space(s) might be approached "otherwise."

Mexican Space Milieux

The first rocket launches in Mexico took place during the International Geophysical Year (IGY) of 1957, but eventually rockets were replaced by telecommunications satellites as the country's favored space technology, as we will see in chapter 2. However, they seem to be making a comeback. The Mexican Encounter of Experimental Rocket Engineering (ENMICE) was originally to take place in Acapulco in November 2023, but the devastation caused by Hurricane Otis forced its organizers to reschedule. So, in April 2024, having been invited to evaluate the social and environmental impact of the students' rocketry projects, I found myself in the mountainous border state of Chihuahua.

The first two days of the competition were held on the urban campus of the state's Autonomous University. In seminar rooms and the hangar of the engineering department, judges heard the teams' project presentations. They examined their rockets to ensure their compliance with technical and safety protocols. At the same time, the public was invited to hear speakers talk about space topics, including the history of Mexican rocketry, collaborations between the Mexican Space Agency (AEM) and NASA, practical applications of space technology, opportunities in space business, innovations in space law, and the anthropology and archaeology of outer space. My fellow judges and I were invited to tour the Center for Training in Advanced Technology (CENALTEC), which shares one of Chihuahua's industrial parks with transnational aerospace companies such as Honeywell and Bombardier. The center's administrators emphasized Chihuahua's positioning as a haven for aerospace nearshoring and its recognition by the *Financial Times* as an "Aerospace City of the Future" offering foreign investment opportunities (Mullan 2018).[1] That evening, ENMICE arranged for us to visit what had been Pancho Villa's home, now a museum run by the federal Secretary of Armed Forces (SEDENA), where we learned about the famed revolutionary's passion for what in the early twentieth century was considered cutting-edge technology: typewriters, telegraphs, and the first small airplanes, which he used to great effect during the Revolution of 1910–1920.

On the morning of the competition's second day, I was eating breakfast in the dining room of the Holiday Inn Express in Chihuahua's capital along with the other judges of the national competition. English was heard everywhere, and, unlike the rest of the country, Mexico's border states had switched to daylight saving time. Two televisions were broadcasting U.S. and world news: on one channel, CNN reporters discussed the violence in Gaza and Ukraine, while on the other Fox News commentators bemoaned the "wokeness" of National Public Radio. "The north is a different world," someone from Mexico City said, referring to Chihuahua and other border states' intimate relations with the United States. One AEM official worried that heightened tensions in Gaza would prevent the administrator of NASA from making a planned official visit to Mexico the next week.

As I drank my café Americano, Pablo, the director of the Federal Civil Aviation Agency (AFAC), showed me a series of wonderfully detailed digital drawings of Marvel superheroes he had done on his iPad.[2] Spider-Man was a particular favorite. The subject of pre-Hispanic astronomy came up, and he quickly sketched a drawing on a napkin of the famous stone relief found in the Palenque tomb of the seventh-century Mayan governor Pakal, known colloquially as "the astronaut" for the motif surrounding Pakal's engraved image that has elements of a spaceship or time machine. Archaeologists usually interpret those elements as a stylized representation of a cosmic tree, but Pablo said he thought the engraving could represent a combustion engine, and I couldn't tell if he was joking. Germán, an engineer and fellow judge, told us about the strange lights and orbs he used to see regularly in the sky in Toluca, and Pablo replied that Germán had probably seen drones or weather balloons. He also mentioned, ominously, a conversation that he had once had with a former member of the Israeli intelligence agency about new technologies that "would completely freak you out." Germán nodded, but he did not seem entirely convinced. The conversation moved on to the relative merits of tortillas made from flour (common in the north) and corn (a staple of Mexican diets everywhere else).

For many of us, the third and last day would be the highlight of the competition, as we would get to see the launch attempts of the student projects that had passed their preliminary revisions. The launches took place in the municipality of Cuauhtémoc, officially the region's "Municipality of the Three Cultures" (not to be confused with Mexico City's "Plaza of the Three Cultures," which we will look at in chapter 5), as the region is inhabited by mestizos ("regular people," according to one interlocutor), Indigenous Rarámuris (Tarahumaras), and Mennonites. For the launches, the organizers of ENMICE had rented a field behind a flight school from members of a local Mennonite community, descendants of immigrants who had originally migrated to Canada from Prussia in the nineteenth century. In the early twentieth century, after undergoing pressure from the Canadian government to conform to secular educational policies, they negotiated with the postrevolutionary Mexican government to establish an agricultural colony in Cuauhtémoc in the wake of the disintegration of the hacienda system. At lunch, I found myself talking to a young Mennonite man who, disrupting common stereotypes about the community's distrust of technology, showed me his cell phone with numerous saved pictures of his new John Deere tractor. However, he seemed to be completely uninterested in space technology. As we waited for the teams to prepare their rockets for launch, "Capi," a local pilot and fellow judge, told me that I had to try the apple pie made in the region with Mennonite produce (and often picked by Rarámuri labor), as well as the famous white Mennonite cheese. Capi lauded the community's commercial success and its importance for the state; however, he also criti-

FIGURE 1 Launching rockets in the Chihuahuan desert. Photo by the author.

cized the Mennonites' resource-intensive agricultural practices, which, given the drought conditions that have long characterized the region, have intensified conflicts with other communities over access to water.[3]

Along with the drought conditions being experienced in Chihuahua, another ecological crisis became a topic of discussion during the EN-MICE rocketry competition, one which, fortuitously, provided me with a way to think about the political, economic, sociocultural, and ecological complications that surround and underlie Mexican engagements with outer space. Oceanographer Mariana and her partner Juan run MAR D' SAL, a company that works with local communities around the world to support the sustainable management of aquatic resources using satellite data. In addition to co-sponsoring the competition, Mariana and Juan gave a talk to students at the University of Chihuahua in which they spoke about their company and its activities to motivate young people to innovate in the space sector while thinking beyond the trope of space exploration. One of their recent projects centers on the rehabilitation of sargasso. Satellite data has enabled the mapping of sargasso's extension over time, leading to a more in-depth understanding of its movements and composition. But their efforts go beyond collecting data. In collaboration with communities that have been affected by the increase of coastal sargasso, MAR D' SAL is looking for ecological and socially sustainable ways to collect, clean, and employ the algae as an organic shield protecting against the erosion of coastal ecosystems, for example, or as an ingredient in locally produced consumer goods.

The hundreds of species of sargasso share a few common characteristics: a golden-brown color, a certain blade morphology, a form of asexual vegetative reproduction, and the presence of small bladders filled with air that allow the algae to float in the open sea. The classical Mayans knew it as *U tail kaknab*, or that which is "thrown by the Lady of the Sea" (Godínez-Ortega et al. 2021, 3). Its common name comes from Christopher Columbus or some unnamed member of his crew, who, struck by the enormous quantities of the macroalgae that occasionally trapped European ships, called this region of the Atlantic the Sargasso Sea. A long succession of navigators and explorers, including Fernández de Oviedo and José de Acosta, described the algae or "sea of herbs" throughout the colonial period.

Alexander von Humboldt, to whom I will return in other chapters, was the first European to pay attention to sargasso not only as a biological genus but as an entire ecosystem that supports a variety of species. He characterized the Sargasso Sea, the only sea whose boundaries are defined by ocean currents rather than land masses, as "a community association, constituted by the algal species and an animal community" (Godínez-Ortega et al. 2021, 4). Indeed, the sargasso community often includes turtles, who use the sargasso mats as a hatching ground for their young, shrimp, crab, eels, fish, and birds, for whom the sargasso milieu serves as a banquet. Sargasso is also considered to be an important species for the capture of organic carbon and thus an actor in the mitigation of global climate change (19). In small quantities, beached sargasso originating in the Sargasso Sea provides food for small coastal species and contributes structural stability for sand dunes, helping to prevent erosion (20). Incidentally, in precolonial America, it was possibly used as a remedy for urinary tract infections (25).

However, the sargasso that has washed up on Mexico's eastern coasts in massive amounts in the last decade is different. Many of these deposits do not come from the Sargasso Sea but from a new concentration of the macroalgae that has recently been discovered in the southern Atlantic tropical zone, termed the Great Atlantic Sargasso Belt. This new "bloom" is likely to have been caused by "human interference with the plant's biogeochemical cycles" and resulting climate change (Godínez-Ortega et al. 2021, 20), including increasing nutrient discharge from rivers into which industrial waste has been dumped. The massive amounts of sargasso that have been deposited on Mexico's beaches originating from the new sargasso belt have affected coastal ecosystems, as the decaying plants reduce the light and oxygen available and release heavy metals such as arsenic. The quantity of sargasso covering the beaches and the extremely unpleasant smell of rotting algae have also affected tourism, which is probably the issue that most concerns Mexican authorities.

In addition to illustrating one of the applications of space technology in contemporary Mexico, sargasso traces a path toward this book's central conceptual and methodological metaphor: the milieu. As a milieu, sargasso is more than a biological agent affecting tourist revenue, as hotel operators in Cancún might argue, but a dense network of relations. It is a social, historical, and ecological assemblage that links the human and the nonhuman; the technological and the biological; the marine, the terrestrial, and the orbital. Its interactions take place at a variety of scales, human and otherwise, from the exchanges between chemical elements and cells that occur at a microscopic level, to the planetary entanglements of ocean currents and biomass, to the extraplanetary satellite gaze. In its messy connections, we can make out traces of colonial desires, apocalyptic warnings, Indigenous knowledge, and technoscientific aspirations.

The Social Studies of Outer Space

The recent boom in studies of outer space in social and humanistic fields makes it impossible to cite a comprehensive bibliography here. I point the reader who wishes to learn more about this blossoming field to Stefan Helmreich's foreword to the 2017 special issue on outer space in the journal *Environmental Humanities* and Valerie Olson's "Refielding in More-Than-Terran Spaces" in the 2023 *Routledge Handbook of Social Studies of Outer Space*. The works cited by these authors include topics that range from organizational sociology applied to national space agencies (Zabusky 1995), to scientific study of the meanings of life (Helmreich 2006), to musings on the implications of testimonies of UFO sightings and alien abductions (Battaglia 2005; Lepselter 2016).

Within this cornucopia of social scientific contributions to the study of outer space, two key themes were fundamental in the formulation of the questions that frame my work. The first is derived from working with "space people" outside the hegemonic centers of space activity. As many authors have pointed out, the discourse on the human exploration of outer space tends to replicate colonialist and imperialist ways of talking about, picturing, and otherwise imagining what humans have done or may do on extraterrestrial worlds. References to "the final frontier" and the need to "colonize Mars" are echoed in the naming practices of space-faring vehicles with references to European explorers from the fifteenth century onwards. Some U.S. proponents of space exploration speak explicitly of their nation's "manifest destiny in the stars" (Rubenstein 2022, 65), a claim that is mirrored in the visual representations of outer spaces as uncanny doubles of the imagined landscapes of the western United States during the nineteenth century (Kessler 2012; Sage 2008). But looking at outer space from Mexico gives a different slant on the colonialist rhetoric that often justifies human activities off-Earth. What is now Mexico came into being after the Spanish conquest of 1521 that began a three-hundred-year-long project resulting in the loss of between 50 and 90 percent of the native population through violence and the spread of pandemic disease (McCaa 2000, 257). In the nineteenth century, the newly independent nation was forced to cede more than half its territory to the United States after the Mexican-American War, one of the consequences of the original policy of manifest destiny, a doctrine that historian Orlando Martínez calls "one of the best euphemisms for bumptious expansionism ever minted" (in King 2000, 64). Both historical traumas continue to play out in everyday discourse, nationalist rhetoric, and popular culture in Mexico; both are present in the outer-space milieux with which I will engage in this book.

I share my concern about the neocolonialist rhetoric of outer space with authors such as Natalie Treviño (2023, 226), who criticizes the space industry for its focus on "capitalist accumulation through colonial activity, language, and logic," and Anishinaabe scholar Deondre Smiles

(2020), who points to the settler logic that underpins much of the discourse on space exploration, as well as the authors of a recent issue of the *American Indian Culture and Research Journal* who examine the complicated relationships between Indigenous and non-Indigenous space science (Shorter and TallBear 2021). The collective Bawaka Country et al. (2020, 2) demands that "promotors of space colonization" pay attention to the multiplicity of worlds that connect with "what Western sciences call 'outer space,'" a call to which I will return in several of the chapters that follow. I owe a particular intellectual debt to the few texts written about outer space from Latin America, particularly the pioneering work of Peter Redfield (2000) on the European launch site in French Guiana, a site that has recently been restudied by Karlijn Korpershoek (2023), and Sean T. Mitchell's (2017) book on the impact of the Brazilian launch site of Alcântara on surrounding communities. In addition, Anna Szolucha's (2023) ongoing research at the SpaceX compound in Boca Chica, Texas (built on land taken from Mexico after the Mexican-American War), takes up many of the themes with which I engage in this book, especially regarding borders on and off Earth. Complementarily, texts by William Lempert (2014) and Juan Francisco Salazar (2017) have inspired me to think about the power of decolonial speculation to imagine alternative futures, terrestrial and otherwise, a topic I will address in chapter 5.

However, despite the vitally important contributions to the social studies of outer space made by post- and decolonial scholars, the experiences of my interlocutors in Mexico point to a more nuanced and ambivalent relationship with the promises of Western hegemonic space discourse. Participants in space milieux are just as likely to unironically advocate for a "Mexican conquest of space" as they are to condemn U.S. corporate greed and national chauvinism. This has led me to emphasize concepts such as *nepantla,* a word that Chicana writer Gloria Anzaldúa (2002, 549) borrowed from Nahua epistemology to characterize the potentials and limitations of the subject who is intersected by borders, simultaneously "both and" and "neither nor" that generate a double vision. Nepantla is also related to the way in which Bolivar de Echeverría (2019) deploys the notion of the "baroque ethos," which the Ecuadorian (naturalized Mexican) philosopher uses to describe the theatrical, practical-aesthetic response to modernity that, he argues, characterizes Mexican popular culture. I will return to these concepts in the next chapter.

The second strand of writing about outer space from the social sciences that I take up in this book concerns the engagements between geographical or spatial concepts and cosmographical or outer-spatial concepts. A key text for me, and one of the first that I encountered when I became interested in outer space, was Lisa Messeri's *Placing Outer Space* (2016), which looked at the ways in which U.S. planetary scientists turn "outer spaces" into "outer places" through toponymy, cartography, analog missions, and other practices. "Place making at a planetary scale," argues Messeri (2016, 12), makes what may be perceived as a vast, homogenous, and alien "outer space" into more familiar, differentiated, potentially experiential locations "that can be navigated, whose dynamics can be observed, and from which lessons can be learned." After reading her work, I began to think about whether there were "Mexican places" in outer space, as well. My doctoral research had revolved around the interactions between historical narrative and place-making practices in the southern Mexican state of Guerrero, where I remember asking about the origins of one municipality's toponym, expecting to hear the story of a local hero who had fought during the war for independence. However, Zumpango de Neri was not named after an insurgent, I was told, but an astronaut. Rodolfo Neri Vela, born in Chilpancingo, Guerrero, spent 9.6 days orbiting Earth on the space shuttle Atlantis as a payload specialist in 1985 and is said to have introduced corn tortillas to the astronauts' diet. "Guerrero has been to outer space," my interlocutors said with pride. Ironically, when I finally met Neri Vela in 2023, he regretfully informed me that the municipality was named not for him but for a distant relative, a nineteenth-century jurist named Eduardo Neri. "The Neri family has many branches."[4]

For some of Messeri's (2016, 31) interlocutors, narrative practices may turn outer "places" into what she terms "place-worlds," which are imagined not only as places you could potentially visit, but places you could inhabit, and which are experienced as both spatial and temporal, existing simultaneously in the past and in the future. Thinking of outer space in terms of place-worlds helps challenge the notion of space as an empty frontier awaiting human expansion. Messeri's discussion of worlds also extends the work of Anna Tsing (2015, 21) on how "world-making projects" allow human and nonhuman beings to create and sustain their lived environments and of Kathleen Stewart (2011) on the need to pay attention to "atmospheric attunements" and ordinary effects in worlding processes. Debbora Battaglia (2014) calls for a "cosmic diplomacy" open to a variety of world-making practices. Similarly, Marisol de la Cadena's (2015) research on "earth beings" and Indigenous worlding is a topic that has enormous potential for rethinking ethical frameworks for space exploration. I came across many worlding practices in my fieldwork, including the carrying out of Latin American analog missions, Indigenous rituals meant to restore or maintain cosmic balance, the narrative construction of speculative Mexican settlements on Mars, and ways of inhabiting the night in rural ejidos.

Valerie Olson's (2018, 6) work on NASA engineers' conceptualization of outer space as an extreme environment in which ecological and technological systems are located provided another way of thinking spatially about conceptual and material relations. Olson writes of the ways in which space systems, like the International Space Station, work to extend and connect natural and geopolitical environments, challenging scalar perception by "unearthing" environmental politics (Olson and Messeri 2015). Her work disrupts linear narratives of space exploration by showing how deeply entangled humans are with space environments. The notion of outer space as part of the terrestrial environment also finds an echo in discussions about the technological objects that circle the planet as elements of an "orbital technosphere" (Gärdebo, Marzecova, and Knowles 2017) and the need to think "vertically" about geopolitics (Graham 2016). Environment as a part of systems thinking, whether technical, political, or economic, is certainly part of the discourse of many of my interlocutors, as I discuss in chapter 2, where I take up the experiences of satellite engineers, and in chapter 3, where I discuss the "commercial space ecosystem" to which many of my interlocutors aspire.

Finally, the "planetary turn" has provided a series of useful provocations for rethinking relational spatiality, showing the embeddedness of the local in global and extraglobal systems and processes. Like Olson's conceptualization of environments, the planetary questions simplistic scalar dynamics and challenges facile divisions between nature and culture. Some of the key authors developing ideas around the planetary, a term first introduced into the social sciences and humanities by Gyatri Spivak (2003), include Dipesh Chakrabarty (2021), whose work on climate change and historical thinking has been crucial in theorizing the relationship between human and planetary scales of time, and Bruno Latour (2018), who argues that thinking from the "critical zone," the thin layer of Earth's surface where most life processes occur, will allow us to move beyond pure localism and abstract globalism. The writings of Anna Tsing and her collaborators have been particularly useful to me for thinking about "frictions" (Tsing 2011, 4) and "patchiness" (Tsing 2015, 4), showing how global processes like capitalism and climate change manifest through specific local assemblages in complex, unequal, and often contradictory ways. And geographer Nigel Clark and sociologist Bronislaw Szerszynski add another layer of complexity to planetary thought by insisting on the interplay between what they call "planetary multiplicity" and "earthly multitudes." Clark and Szerszynski (2022, 76) argue that the planet itself is "inherently multiple and self-differentiating across a range of scales" and interacts with a diversity of human collectivities who "organize their everyday material practices in response to the challenges and opportunities arising from the Earth's inherent multiplicity." Mexican satellite engineers interested in Earth observation and monitoring share a preoccupation for the *planetary* with artists critically interrogating the relation of humanity to the cosmos and local agricultural communities suffering from the effects of climate change and global capitalism.

Indeed, each of these terms—place, world, environment, and the planetary—speaks to some aspects of the matters of concern (Latour 2004) or care (Haraway 2016) with which I and my interlocutors engage. And related terms, like cosmos, nation, landscape, territory, and border, also appear in these chapters. So why add another concept to this conceptual density? Milieux, I argue, does the work that none of these words taken separately can perform. Places and worlds signal affect and experience, while environments and the planetary emphasize connection and friction. Milieux are scalar, active media in which and through which a variety of life-forms emerge, interact, and are mutually transformed. They combine affect and relationality, simultaneously taking into account the intimacies of everyday life and its multiple and multidimensional extensions in time and space. They entangle and multiply places and worlds both internally and externally, while insisting on the material effects of planetary environmental relationality.

Thinking With and Through Milieux

The concept of milieu came to the humanities and social sciences from biology largely through the work of philosopher and medical doctor Georges Canguilhem (2008, 90), who called for paying attention to the complex and generative relations between humans and their environments (or milieux), defining life as "an attempt in all directions." The word has recently regained popularity in the social sciences, as evidenced in a 2019 special issue of *The History of Anthropology Newsletter* dedicated to exploring the applicability of Canguilhem's milieu to anthropology. In her contribution to the newsletter, Kathleen Stewart (2019) describes milieux in terms of "animus and impact," calling our attention to

> the prismatic singularities of an actual scene of composition and decomposition forged in fractious points of contact that inspire and directly induce lines of action or simple shifts in direction or duration . . . an atmosphere with qualities, an imperative demanding a response, an objective, a pooling up that can overflow its bounds, a track on which to somehow venture out.

FIGURE 2 Sargasso on the beach. Photo by Paulina Zanela / MAR D' SAL.

A milieu is both singular and multiple, as it is defined by the speci-
ficities of concrete interactions but is constituted by diverse elements in
relation to one another. In this book, I use the plural "milieux," whose
x insists not only on the milieu's internal multiplicity, but also on the
broader diversity of these "scenes of composition and decomposition."
Just as there is no singular "outer space," there is no singular "outer-space
milieu," although the various milieux rub against each other in complex
ways, as we saw in the case of the Chihuahuan rocket competition.

Milieux enrich spatial conceptualizations such as place, environment,
and ecosystem in key ways. They imply a dynamic relationality that goes
beyond the social construction of space by emphasizing the ongoing, re-
ciprocal relationships between inhabitants and their surroundings. People make places meaningful, but places also shape human possibilities
and forms of life. The example of sargasso hints at a second implication
of thinking through milieux: the interweaving of biological and social
processes and the relations between human and nonhuman actors.[5]
Humans interact with satellites and robots that allow for observation,
communication, extraction of resources and data, and, for some, the poetic, prosthetic extension of the human into the cosmos. They also form
relationships with nonorganic materials like stone pyramids, moon dust,
geological formations, and meteorites; with biological actors such as sargasso, nocturnal animals whose circadian rhythms have been disturbed
by light pollution, and marine fauna now being tracked by satellite; as
well as with hard-to-classify alien beings.

Stewart's attention to the atmospheric qualities of milieux suggests
a third dimension that concerns us here: their involvement in the affective dimensions that are crucial to lived experience. My interlocutors are
bound to outer space and each other through common passions, uncomfortable frictions, hopes, expectations, and deeply rooted fears.

Finally, the concept of milieux emphasizes process and becoming rather
than fixed states. As I show in the following chapters, from references
to pre-Hispanic astronomical practices to planned or imagined Mexican
futures in space, the temporal dynamics of outer-space engagements in
Mexico were evident in almost every interaction. History and collective
memory, understood as cosmohistorical narratives, shape Mexican space
milieux in profound ways, complicating notions of linear notions of prog-

ress and development by insisting on a constant *ir y venir* between past, present, and future.

The word *milieu*, Stewart reminds us, orients us toward both a *medium*—the "surround" that allows for exchange—and a *middle*. It challenges the verticality often invoked in outer-space discourse, preferring instead a multidimensional spatial extension with blurry boundaries that I think of as a kind of connected situatedness. In this same vein, Isabelle Stengers (Stengers, Massumi, and Manning 2009) calls for a "mesopolitics" that allows for "thinking through the middle," avoiding the traps of the excessively detailed micro and the unduly abstract macro. A meso-scaled milieu is a site for reflecting on "how the relations between the micro and the macro are assembled. . . . It's about everything that the macro does not allow to be said and everything that the micro does not permit to be deduced." It is the scale of narrative rather than deduction. There is no center to a milieu, but rather of sense of being "in the midst" of things, which seems particularly appropriate for an anthropology undertaken from "inside" social phenomena.

Thinking from the middle seems particularly appropriate for a study centered on Mexico, in what is sometimes referred to as "Meso" or Middle America.[6] Indeed, there is a sense that Mexico is caught somehow in the middle, so far from God, they say, so close to the United States,[7] whose gravitational pull is difficult to resist, entangled as it is in webs of power and desire. The nation-state and its borders are extremely visible, in counterpoint to the flows of people, objects, and ideas that have always characterized Mexico. The nation has been both victim and perpetrator of colonialist violence, a beacon of international peace and the site of bloody state repression. Discursively founded on the notion of *mestizaje*, a word which also implies a "middle" and refers to the mixing of Indigenous and European identities, nationalist history pays homage to its pre-Hispanic roots while marginalizing contemporary Indigenous populations. Here, rocket ships and satellites share imagined outer spaces with caravels, feathered serpents, and pyramids. In one of the narratives produced by my friends in the collective Marsarchive.org (see chapter 5), astrocapitalist Elon Musk and Aztec war deity Huitzilopochtli both have a presence on Mars. Satellites are named AzTechSat, and Moon robots are *tepoztli* ("metal things" in Nahuatl) or jaguars (see chapter 4). "We have a right to claim a Mexican place in outer space," many say, "because we have always been there." At the same time, other interlocutors express a fear of being left out of a modernity that increasingly depends on extraplanetary technology. Outer space has thus served as a not-entirely-blank canvas for the projection of an array of desires and speculations, including nationalist and neocolonialist futural aspirations, but also, as we shall see, speculative proposals that are explicitly decolonial.

This book echoes Stengers's call for a mesopolitical cosmopolitics, one that not only narrates the cosmohistories that emerge in Mexican space milieux, but also opens a space for collectively thinking through matters of common concern, for questioning the decisions for futures off and on Earth made in the name of "humanity," for reclaiming the right to have a voice in those decisions, for taking advantage of milieux's stickiness, slowing down the "race" to "conquer" space, and, in the process, turning "outer space" into "outer spaces." The notion of milieux becomes a starting point for examining how (human-driven) geopolitical and (more-than-human) planetary-scale processes are entangled in daily practices of both global and local worldmaking. More than mere settings or contexts, milieux organize modes of attention and objects of attention and direct the kinds of questions that can be asked and the kinds of relations that can be traced. They make no necessary distinction between nature and culture, past and future, or outer and inner space. Thinking with and through milieux has allowed me to orient my study toward the kinds of material and affective relationships space people have with each other, with objects and ideas, relationships that come into being through scalar tensions between the cosmos and Earth, Mexico and the rest of the world, the past and the future. They permit me to both expand my attention beyond the minutiae of microenvironments, like agencies and laboratories, and ground seductive but abstract cosmic questions.

In methodological terms, interacting with my interlocutors requires following them as they move between spaces, physical and virtual, being flexible enough to let "fieldwork" emerge in the interstices of everyday life, rather than existing as a framed time and place set apart and demanding just as much "participation" as "observation." I have become part of the milieux I study, just another "space person." In my experience, participating collaboratively with diverse actors has been a way to access ethnographic information while "making myself useful." George Marcus's notion of ethnographic "para-sites" has been a helpful way to conceive of the spaces of collaboration. Marcus (2000, 2) conceives of these spaces as encounters between "counterparts" (rather than the "others" whom anthropologists often perceive as their interlocutors), that is, "coproducers of interpretations that we elicit, cajole, contest or share." Each encounter in a para-site "decenters the conventional ethnographer-informant relation through a para-ethnographic epistemic partnership with expert interlocutors or by involving audiences in projects of media and knowledge making" (Boyer and Marcus 2020, 13). My interlocutors have expertise in robotics, computer programming, space law, design, and artistic production. One of the benefits of working with experts from different disciplines has been learning from them and generating productive dialogues around questions like "Will Mexico's first Moon mission be considered space heritage?", or "How do dark skies impact the experience of gendered violence?", or "Is space really for everyone?"

Some collaborations have involved conferences about the sociocultural aspects of outer space ("human factors," in emic parlance); this was the case for a series of meetings with the Colombian and Mexican chapters of the SGAC in which we discussed what a Latin American "outer-space culture" might entail, and how to rethink ethical space practices (see chapter 3). Others have centered on practical implications of critical outer-space thinking that considers gender equality, cultural diversity, and sustainability in the solutions to problems, like my participation in the ENMICE competition described above. Other encounters focused on interacting with nonspecialist publics through speculative workshops for the imagining of diverse outer-space scenarios; examples include "futuring" workshops that I helped facilitate at the AEM's headquarters in Mexico City, with a potential dark-sky park in Hidalgo (see chapter 6), and with the collective Marsarchive.org, which regularly organizes ludic encounters that provide a space for the science-fictional imagining of a future-past Mexican presence on Mars (see chapter 5). Once the province of literary and film science fiction, speculation has become a potent strategy in industry, design, and business for tracing social and economic trends, as well as projecting possible futures. The anthropological use of speculation adds a critical edge to these projects by calling for speculative fiction to "become not only a resource for imagining alternative worlds but also a medium for remaking our presence in the world" (Anderson et al. 2018).

The more I study how Mexicans engage with outer space, the more I realize how much these engagements reveal things that are happening on the ground in Mexico. Looking at projects undertaken at the Mexican Space Agency, for example, has provided a window into, among other things, the functioning (and misfunctioning) of the Mexican government, the negotiation of political agreements, the flow or diversion of money, the interactions between the public and private sectors, the advancement of government workers' career trajectories, the maintenance or contestation of international relations, and the construction of public discourse. Working with scientists at the National Autonomous University of Mexico (UNAM) has helped me understand some of the same issues, but in an academic setting that, its paradigm of scientific objectivity notwithstanding, operates within the context of university and national politics, and interpersonal, institutional, and international collaborations and rivalries. Discussions with young people in STEM fields who consider themselves part of the "Space Generation" have helped me understand their mix of optimism and frustration with the status of the space sector in Mexico as well as their dreams for the future, both individual and collective.

Talking to artists and observing their work has helped me understand how it might be possible to use space exploration as a pretext for reflecting on alternative Mexican futures that, without abandoning the ludic possibilities of space imaginaries, foreground pressing local and national issues like immigration, feminicide, environmental degradation, and the historical legacies of colonialism. Interactions with rural community leaders interested in promoting astrotourism have revealed the complications of community organization and collaboration with national and international institutions, and local perspectives on identity, history, and relations with the environment, as well as how economic conditions and survival strategies have been transformed in the twentieth and twenty-first centuries. Finally, informal talks with people who do not participate in space activities have shed light on how ordinary Mexicans do or do not think about outer space in their everyday lives.

Geographically, my interlocutors occupy particular spaces: home, community, laboratory, or office, as they evoke senses of belonging to institutions, states, Mexico as a nation, and humanity as a whole. Often, they cross borders as part of the activities that tie them to a cosmopolitan "space community" that, for all its pretensions to inclusivity, rests on practices of geographic, ethnic, and gendered exclusion. "I wanted to be an astronaut," one of my interlocutors told me, and she "checked off almost everything on NASA's list"; that is, until she came to the last item: "Must be a U.S. citizen." But for many of them the inclusive potential of the cosmos triumphs. "Space *is* for everyone," they repeat, "or at least it should be."

This Book

The book is divided into six chapters that tell separate but related stories about Mexican space milieux. I start with "Cosmohistories," a look into the long history of encounters between populations of what is now called Mexico and the cosmologies, astronomies, cultural discourses, and practices that make up much of the texture of Mexican space milieux. The chapter opens with a description of the postrevolutionary aesthetic construction of *la raza cósmica*, the cosmic race, and mestizaje as the basis for national identity. Through the common phrase "We have always looked to the stars," I reflect on some of the features of precolonial cosmology, based in part on experiences of observation of the horizon and expressed in architecture, pictorial writing, and ritual practice; I then turn to the cosmological orientations of colonial Mexican intellectuals in an attempt to understand both the relations between colonial and precolonial discourses and the relations between colonial and European scientific knowledge and practice. I conclude the chapter with the mesopolitical Nahua concept of nepantla, a liminal, baroque positioning that will reappear later in the book.

Chapter 2 moves from the cosmic race to the space race. Even as they also say that Mexico has always had a relationship with space, many of my interlocutors also insist that, really, it all began with Sputnik. After a brief discussion of the early history of Mexico in space, I look at how the country became both "globalized" and "planetized" in the wake of the orbitalization of telecommunications after the launch of the Morelos satellites (and their accompanying astronaut) in the 1980s, while institutional discourse simultaneously insisted on the importance of national identity and technological sovereignty. The Mexican government would incorporate the orbitalization of technology into a teleological narrative that projected the development of "Mexican satellites" as both the logical evolution of pre-Hispanic astronomical knowledge and the means through which the nation would become fully modern. This chapter traces the history of the sociotechnical imaginaries (Jasanoff 2015) of the institutional space milieu in Mexico, which are marked by a series of spatiotemporal tensions between nationalism and geopolitics, aspirational futures and structural exclusions.

The third chapter, "The Space Generation," centers on the implications of *borders* and *generations* as concepts that cut across the universalist pretensions of outer-space discourse. Here, I take up the experiences of "Mexican astronauts" who have made their mark in the milieu of the Mexican space sector, reflecting on their experiences of the tensions between the view of a shared, borderless Earth seen from above and a terrestrial world intersected by different kinds of borders. I then trace the futural aspirations and frustrations of younger participants in Mexican space milieux by describing my interactions with members of the Space Generation Advisory Council (SGAC), an international organization made up of students and young space professionals, many of whom are excluded from full participation in the cosmopolitan "space community," even as they express their passion for outer space in creative ways.

Chapter 4, "In the Navel of the Moon," expands the notion of space milieux to include nonhuman, planetary actors and returns to the possibilities of cosmic diplomacy at different scales. I focus on Mexico's relation to the Moon, an important celestial body for pre-Hispanic Mesoamerican populations, but also for modern technoscientific imaginings. This chapter's discussion of the contradictory ways in which the Moon interacts with humans and humans with the Moon leads to a series of reflections on the worlding practices that pluralize "space" and question its status as "outer." I look at the Apollo Moon landings seen from Mexico and reflect on Mexico's participation as a signatory to a series of international space agreements, from the 1967 Outer Space Treaty to the NASA-led Artemis Accords, signed by Mexico in 2021. Through an analysis of recent and near-future projects developed by Mexican university laboratories and start-up ventures meant to send robots to the Moon, particularly the 2024 launch of the "Colmena" project, I examine national aspirations for the creation of a "space ecosystem" and the Mexican space sector's complex, colonially mimetic relation to the U.S.-based NewSpace industry. After discussing the meanings of "failure" and "success" in the context of Mexican space activities, I turn in the last section of the chapter to the debates around the Moon's potential status as the site of resources to be exploited, a heritage site, a location for pristine scientific experimentation, and a "person" in legal and cosmic terms.

The fifth chapter, "Transhabiting Mars," continues the book's journey through the solar system toward Mars, a planet that condenses technoscientific and artistic temporal imaginings. The chapter revolves around the cultural collective Marsarchive.org and its ludic "transhabitation" (Olson 2018, 143) of the red planet as a means of deterritorializing and reterritorializing, dehistoricizing and rehistoricizing, Mexican space milieux. I look at several versions of the collective's collaborative workshops, each taking as its starting point a preconquest city located in Mexico's central valley as a *lieu de mémoire,* or site of memory (Nora 1989), as it is repurposed as part of a speculative project about Mexican futures in outer space. Martenochtitlan imagines the first Mexican city on Mars and becomes a way of thinking about mestizaje and hybridity, returning to the "cosmic race" I first discussed in chapter 1. Martelolco, Martenochtitlan's twin city, becomes the site of Mexican memory, a multilayered temporal palimpsest of precolonial magnificence, violent conquest, modernist aspirations, international diplomacy, and national tragedy. Finally, "Mars in Guerrero," a truncated version of the collective's activities and the only workshop that did not start from the nation's center, sets forth a series of precepts for a Mexican social utopia. This chapter, which focuses on popular culture's potential to disrupt the aspirations of technoscience, takes up the possibilities of speculative narration in a baroque key.

"Dark Skies," a return to Earth and to the starry cosmos experienced from below, closes the book. In an age of constant illumination from unshielded factory city lights, electronic glare, and a growing number of satellites, the night sky has become, in the words of some of my interlocutors, a finite resource. In this last chapter, I describe a series of efforts by one community to reclaim its night sky through astrotourism, a complement to the ecotourism that has largely replaced the historical economic activities of mining, agriculture, and immigration. Reflecting on the anticolonial possibilities of "darkness," I end by circling back to a notion of milieux as situated cosmos: relation, landscape, and multiscalar immersion that combines the geo and the astro, the human and the eco, confounding easy distinctions between up and down, here and there.

In between the chapters, I have included a series of "Interludes" that take up artistic and cultural engagements between Mexico and outer space that allows me to reflect on these entanglements from a more poetic perspective, deepening and, perhaps, confounding the arguments presented in each chapter.

CHAPTER 1

COSMOHISTORIES

I start this chapter with a series of testimonies recorded in planetariums in Mexico City as one path into the stories I want to narrate about Mexican encounters with outer space. I could have started at the very beginning, with the creation of the first sun, when the gods sacrificed themselves so that spacetime could unfold. Or I could have started with a different beginning, like the launch of the first rocket from Mexican soil into the stratosphere. However, starting this story from the middle (the milieu) and moving back and forth in time allows me to trace nonlinear trajectories, generating what Federico Navarrete has termed cosmohistories. Instead of a "monohistorical" regimen, Navarrete (2021, 25) argues,

> we should erect a truly cosmohistorical regimen that recognizes the existence of diverse historicities, or rather different historical worlds that produce different chronotopes, include diverse protagonists beyond the human, and conceive of distinct forms of the historical process. Like the cosmopolitics that inspired it (Stengers 2005), cosmohistory does not try to construct a singular truth or a unified historical world; rather, it concerns itself with the comprehension of the always complex, violent, and fragile interactions between the historical worlds whose totality is unknown and perhaps unreachable, in order to construct partial and negotiable historical truths.[1]

The cosmohistories I present in this chapter are certainly partial and negotiable. They focus on how precolonial Mesoamericans conceived of and acted upon the fabric of spacetime, but they also take up the colonial processes of cosmic reconfiguration framed by contemporary non-Indigenous discourses about "ancestral" astronomical knowledge and practice that are often elements of the cultural repertoire of participants in Mexican space milieux.

The Cosmic Race

In June 2019, I asked my students Rosa Inés Padilla and Mariel Carpio to interview visitors to several of Mexico City's planetariums about their views on outer space as a counterpoint to the perspectives of interlocutors in the Mexican space industry.[2] They talked to visitors to the capital's oldest planetarium at the National Polytechnic Institute (IPN), which opened in 1967 and was named for astronomer Luis Enrique Erro, at the Universum science museum at the UNAM, and at smaller planetariums dotted around the city. Interviewees talked about their first memories of outer space, encounters with telescopes, reading Isaac Asimov and Ray Bradbury, subscribing to the popular science magazine published by the Institute of Astronomy in Tonantzintla, Puebla, becoming fascinated by astrology, and gazing at the night sky.

Rosa Inés and Mariel also visited the Tunnel of Science inside La Raza, a subway station in north-central Mexico City. Installed in 1988, the Tunnel's most impressive feature is the Celestial Vault, a long, dark hallway illuminated only by phosphorescent paintings and lights that represent the stellar constellations. It also houses a permanent exhibition of astronomical photographs and temporary displays related to science and technology. Rosa Inés and Mariel talked to a few representatives of the more than two million commuters who pass through every month as they transfer between lines three and five of the metro system. Students, farmers, government employees, housewives, and retirees talked about cosmic awe and curiosity, how much money the United States is investing in outer space, the existence of life on other planets, strange lights they've seen in the sky, the lack of government interest in science and technology, whether NASA really went to the Moon in 1969, their favorite science fiction movies, memories of dark skies, ancient pyramids, and ecological crises.

"The thing is," said Rey Humberto, an organic farmer in his sixties, "we're just a tiny part of the universe. We're nothing."

> The universe is so big, and there's so much to discover. And sometimes you just think, I guess like the astronauts and everyone else, "Well, where do we come from? What are we doing here? Why are we here?" And *híjole*, no! I mean, you look and say, "No, when will I go to the Moon?" I don't think we will ever have the opportunity to see all of those wonders, no? Or maybe, I think, in the future. But, well, it isn't for everyone. Just the ones who have money will be able to travel because it's so expensive.

He loves looking at the stars at night, although you can't really see many in the city. He has also seen UFOs.

> Who knows what happens if they take us? I wouldn't enjoy that, or maybe I would if I wasn't forced. . . . But I like to watch the stars at night and see how [the UFOs] fly by. At such speed! So that you tell yourself, "Hey, our . . . technology can't achieve those speeds." No, truly, there is life on other planets. Probably nicer, although they've probably made mistakes, too, like us. Because of all this, now everything to do with the planet is taking us toward chaos. We're doing ourselves in. And now there's no return. This is coming, and it's coming for everyone. We've destroyed our planet; there are too many of us. And everything to do with business is about making more and more money.

Rosa Inés asked Irene, a government worker, if she thought humans would populate outer space someday.

> Well, exploration, yes, I think so. Because for better or worse, we hunger for knowledge of new things. But to populate . . . well, I think that if there are more people out there . . . well, it would be like an invasion, no? Because it would be like if they came to populate the Earth. I mean, like, every planet would have its own people and that would be an invasion. Yes, an invasion of their space, their lives. That's what I think.

Gerardo, a retired accountant, was skeptical: "I don't know how important space is, compared to what we're doing down here, everything we're
destroying, no? Why should we go thousands of meters up? When we
live here, down below?"

The metro station La Raza is named for the nearby Monument to La
Raza, built between 1930 and 1940 to highlight the pre-Hispanic elements of Mexican identity. Engineer Francisco Borbolla and architect
Luis Lelo de Larrea collaborated in the monument's design: a pyramid,
with an eagle devouring a serpent on its pinnacle, and a series of statues
at the base that represent the founding of Mexico-Tenochtitlan and its
defense against the Spanish invaders. Serpent heads in the style of Teotihuacán adorn the pyramid's steps. At the same time, engravings refer to
the pre-Hispanic Triple Alliance, the last Mexica *tlatoani* (ruler) Cuauhtémoc, and the plumed serpents of Xochicalco. Today, bright blue and
red graffiti add unintended decorative elements. The icon that marks the
location of the metro station on the subway map is a simplified drawing
of the monument: an eagle sitting on the top of a pyramid. Although
the belief in a shared Hispanic history and identity condensed in the
phrase "the Ibero-American race" had begun to be promulgated from
the middle of the nineteenth century, the "*raza*" of the monument and
metro station refers more specifically to *la raza cósmica*, or the cosmic
race, promulgated by philosopher and educator José Vasconcelos in 1925,
four years after the end of the Mexican Revolution. It would become
one of the founding texts of the postrevolutionary ideology of *mestizaje*.[3]
The raza that Vasconcelos envisioned would have the capacity to build
a utopic new civilization to be called "Universopolis," that would promote the universal expansion of knowledge and morality. This "fifth race"
could only emerge from Ibero-America, given its mestizo biological and
cultural heritage. It would be the task of the state, through *indigenista*
educational programs and social policy (assisted by the new discipline of
anthropology), to acculturate "the Indian," conceived as both a necessary
ingredient for and an obstacle to the cosmic race.

Vasconcelos used his position as Minister of Education to promote
and support the creation of monumental public artworks extolling a
mexicanidad rooted in the virtues of mestizaje and the Revolution,
giving rise to the twentieth-century Mexican muralist movement that
came to be identified as "The Mexican School of Art," whose most

prominent representatives were Diego Rivera, José Clemente Orozco, and David Alfaro Siqueiros, known as "The Big Three," as well Rufino Tamayo, considered by many to be the fourth great Mexican muralist. From 1921 until the end of the 1960s, these and other painters would cover the walls of Mexico City's public buildings with massive landscapes, historical scenes, and allegorical figures meant to educate and inspire the mostly illiterate postrevolutionary populace. However, although they represented notions of mestizaje and social progress in their works, as in Vasconcelos's original vision, muralists broke with their patron in a variety of ways, expressing distinct visions of both "race" and "the cosmic."

Rather than the "spiritually directed racial eugenics" that characterized Vasconcelos's thought, Diego Rivera depicted a "demographically mixed society, with culture as its unifying medium" (Coffey 2012a, 9). Rivera's murals combined a romanticization (and homogenization) of Indigenous culture with a technologically optimist orientation toward the future, portraying workers, peasants, and Indigenous actors as co-constructors of the modern nation. The culture of "deep Mexico," he believed, would serve as the basis for the industrialized, harmonious future. For "heretical Marxist" Rivera, "humanity and the revolution function as cosmic substances in a concrete world," writes Alejandro Anreus. "In Rivera's vision, utopia is inevitable" (Anreus 2012, 47).

Anarchist Orozco, by contrast, eventually produced murals that were often critical of the violence that characterized the Revolution, avoiding the mythologizing nationalism that characterized Rivera's work. Instead, Orozco viewed human history as "a process of constant struggle against betrayal and corruption, where sacrifice is never-ending and redemption is at best a distant promise" (Anreus 2012, 46).

Siqueiros was a self-proclaimed "citizen-artist" and a Stalinist communist who had participated in a failed plot to kill Leon Trotsky three months before Soviet operatives assassinated the exile in Mexico.[4] In print, as well as in works such as the paintings *Ethnography* and *Tropical America* (1932), Siqueiros attacked Rivera's romanticization of Indigenous culture and denounced the marginalization of contemporary Indigenous groups by their governments. The cosmic future, for Siqueiros, was not to be found in the Indigenous cultures of the past, but in the creation of a politically aware proletariat (Folgarait 1987, 34).

In the 1930s, Rufino Tamayo began to paint images of humanity's relation to the cosmos in a philosophical and psychological key, eschewing politics and often referencing pre-Hispanic iconography. Born in Oaxaca to Zapotec parents, Tamayo's interest in Indigenous aesthetic forms only deepened after Vasconcelos hired him as an illustrator at the National Museum of Archaeology, History, and Ethnography. Much of Tamayo's work sought to transmit the "essence" of Indigenous cosmology as "what makes [Mexico] different and at the same time universal" (Torres 2006, 13). Opposed to a conceptualization of mestizaje rooted in cultural and biological fusion, Tamayo defended a national identity based on a plurality of ideas and cultures (Torres 2012). However, like Rivera, Siqueiros, and other muralists, Tamayo would later create works that expressed his ideas about the possibilities and dangers of modern technology. I will return to the muralists' ambivalence about technology in the context of the space race in chapter 4.

The generative forces unleashed by the revolution had dissipated by the mid-twentieth century, as the muralist movement became co-opted by a corrupt, single-party state that quashed dissent through censorship and violence, although it continued to present itself as the standard bearer for revolutionary ideals. That said, muralism continues to be a vital art form today, in many ways returning to the movement's original mandate as a genre for "the people," and a medium for popular expression in a context characterized by the contradictions between "the cultural legacy

FIGURE 3 *Duality*, Rufino Tamayo. Photo by the author.

of revolutionary Mexican nationalism and the increasingly neo-liberal and authoritarian Mexican state, between Mexico's popular sectors and the political apparatus that demanded their obedience" (Campbell 2012, 265). The styles and themes of twenty-first-century Mexican muralism owe as much to popular culture, comics, video games, and Hollywood movies as they do to pre-Hispanic cosmology and official pedagogy. That said, for Chicano muralists, the notion of a "cosmic race" has become a critical concept that performs a "radical mestizaje" in response to a similar (but not identical) set of precarities in the United States (Barnet-Sanchez 2012). And, as we will see in later chapters, contemporary artists who work in other media have questioned official culture's construction of mexicanidad by producing works that deconstruct nationalist myths, by refusing to engage with those myths at all, or by questioning the possibilities of art in a national context characterized by political, economic, and social precarity (Barrios 2022, 10).

Despite historical tensions and contemporary transformations, Mexican muralists' concern for the idea of a modern nation with deep historical roots and their reflections on the tensions between the national and the cosmopolitan motivated "a radical change at the core of the social imaginary" and "a founding order, even at the expenses of historical accuracy" (Reyes, in Coffey 2012b, 60). Thanks to primary school textbooks and other pedagogical materials, civic education in Mexico today continues to incorporate the muralists' imagery to feed a national identity that revolves around normative racial and cultural mestizaje (see Vaughn 1997), while a widespread belief in Mexicans' cosmic inheritance continues to inspire futural imaginings.

"We Have Always Looked to the Stars"

Outside the Luis Enrique Erro Planetarium at the Poli, Iván told Rosa Inés and Mariel that he thought Mexico's involvement in the cosmos began with "our roots." "Since ancient Tenochtitlan, that's where our roots are. They saw the sky and adored the stars, the Sun, the Moon, that's why there are pyramids." Enrique, 75, and his brother Raúl, 65, both engineers, were fascinated by space for its capacity to inspire abstract thought and reflection. Enrique mused, "The pre-Hispanic cosmogonic vision is famous. The temples are really observatories, the way their constructions are oriented to things like the sunrise or the constellations. It's really interesting and impressive." He told my students that they should visit some of Mexico's archaeological sites. "Teotihuacán is the classic one. . . . Its orientation isn't north–south; actually, it's based on Orion . . . the pyramid of Quetzalcóatl, the Pyramid of the Sun, and the Pyramid of the Moon are exactly oriented toward Orion. They were careful observers." Raúl wondered how pre-Hispanic astronomers could be so exact, "without being able to use calculating machines." For Ernesto, a secondary school teacher, space was filled with nostalgia:

> A long time ago, well, since we went to Teotihuacán and climbed the pyramid . . . well, we've never gone at night, because they don't let you, but, well, contemplation: looking at every detail, one day at a time, for a long time. I think it was a really good process for people in the past that didn't have any technology. And now, with the famous satellites and all of that, they sweep away your vision, no? Before, observation was the most important thing. But now, with all the contamination, you can't see anything.

According to Norma, an engineering student in her twenties:

> We talk a lot about how, for example, the Spanish brought us so much culture, literature, and things like that. But, well, here, the Aztecs, the Mayas, already had a formidable culture. For example, numerically they were super advanced, and in astronomy, too. I mean, they had really impressive advances for their time, with the tools they had. And I think that so much of that culture was destroyed that we'll never truly know what we had. . . . I think they had some advantages over us, unfortunately. And, well, you can't really say that with the Spanish we were better or worse off, just that it was a culture clash, no? Obviously, they took a lot of things away from us, but we learned other things, no? So, ultimately, we're a mix of everything, no?

In the planetarium at Universum, the science museum on the UNAM's main campus, neuropsychology student Mariana had just seen the program *Mayan Archeoastronomy* shown as part of the roster of programs at most Mexican planetariums. With audio tracks in Spanish, English, Portuguese, and Chinese, the film was developed by the National Council of Science and Technology (CONACYT).[5] It begins with the deep blue of the nighttime sky. "The darkness that surrounds opens up, and we observe the sky. We discover *Nohek*, Venus." A yellow sphere with faint glyphic tracings comes into view. "*Uj*, the Moon." A slightly larger blue-green sphere floats by. "And from the Earth, we see a constellation, the stalking *Balam*, a jaguar that carries the stars on his skin." The outline of a feline with flaming red eyes appears in the middle of the screen, and the night sky becomes a nocturnal jungle. We see the birth of a ceiba, a Mayan symbol of the cosmic tree that connects the layers of the cosmos. The narrator poetically describes the movement of the sun *Kin*, the emergence of time, and the creation of the Mayan calendars. "Everything ends and then begins again." Animated images illustrate the construction of Mayan temples, like the ones in Chichén Itzá, Bonampak, and Uxmal, which served as both representations of the cosmos and sites for its observation. They show a diagram of the relationship between the two main calendrical cycles, one solar and one ritual, as two interlocked clockwork cogs. The narrator speaks of "our astronomers"; the rain god Chaac, the lunar deity Ixchel, and the nocturnal deity Balam; Orion, Aldebaran, Mars, and the Pleiades; the observation of eclipses; the white rabbit in the Moon; and star glyphs that are also observing eyes. "We are wings that transform the sky's flight. . . . We are like Balam, jaguars that carry the stars on our skin."[6] Rosa Inés asked Mariana how the film made her feel:

> I don't know, it's a kind of pride, a feeling of excitement because it's like we were a bunch of badasses (*chingones*)! I mean, it's like, I don't know, thinking about how the Spanish or other foreigners say that we were savages or that we didn't have any knowledge when it wasn't like that at all. When years later we know how much we got right, no? It's like they could have seen the whole future.

I hear the phrase "We have always looked to the stars" all the time, from diverse participants in Mexican outer-space milieux including astronomers, engineers, artists, and community organizers, as well as members of the general public. Usually, the reference is to pre-Hispanic astronomical practice, particularly the observational prowess of ancient Maya and

Nahua populations who are considered to compose one branch of modern Mexicans' cultural and biological mestizo ancestry. But what are the deeper implications of this phrase? Who is "we"? What does "looking" entail? And what do we mean when we say "stars"? In this section, I look at the construction of this image of precolonial Mesoamerican populations as protoscientific stargazers, ancestors of contemporary astronomers whose knowledge rivaled that of their Eurasian contemporaries, as well as the arguments of scholars who consider that the night sky was only one element of the complex precolonial processes by which the cosmos was spatially and temporally ordered.

Mesoamerican Milieux

Before the arrival of the Spanish in the sixteenth century, what is now Mexico was home to a wide variety of culturally and linguistically diverse populations, who had inhabited the continent for thousands of years. Much of this region comprises what has come to be called Mesoamerica, an area usually defined by archaeologists as including northern Costa Rica, Honduras, El Salvador, Guatemala, Belize, and central and southern Mexico, as well as the regions of northern Mexico that border the Gulf of Mexico to the east and the Pacific Ocean to the west, a geographical *U* shape that is much thicker at the bottom than the sides. However, the concept of Mesoamerica is not based on geography, but on culture and history. Therefore, its limits were never clearly defined, as "Mesoamerican influence" could be found far outside the area's borders and waxed and waned at various times. Taking a page from Alfred Kroeber's notion of "culture areas," anthropologist Paul Kirchhoff proposed the term "Mesoamerica" in 1943 as a productive way of analyzing the commonalities between the preconquest populations of central and southern Mexico and the northern regions of Central America, and of distinguishing these groups from their neighbors who did not share the same political and cultural practices. Kirchhoff (2000) characterized Mesoamerica according to a series of traits, including sedentarism, the cultivation of maize, the practice of the ritual ball game, and the development of similar religious and calendrical systems.

In Mexico, the concept was further developed by Alfredo López Austin (1973; 1980), whose works gave rise to the "cosmovision" school of Mesoamerican studies. López Austin and his colleagues contributed to an image of a relatively unified symbolic worldview shared by a broad geographical and linguistic range of populations, whose aspects included an underlying dualist logic that opposed day and night, up and down, life and death, male and female, cold and hot. Mesoamerican cosmology also incorporated a tripartite vertical cosmic structure often represented as multiple layers above and below the human world; a quadripartite horizontal structure organized around the sides or quadrants of (time) space, rather than cardinal points; and, finally, a 260-day ritual calendar. Although these elements can be found in a variety of sites and contexts, it is important to note that the populations of Mesoamerica were internally stratified as well as ethnically and linguistically diverse. They included speakers of Mayan, Uto-Aztecan, Otomanguean, Mixe-Zoque, Totonacan, and several isolated languages, such as Purépecha, Oaxacan Chontal, Xican, and Cuitlatec. The history of Mesoamerica was already one of intense interaction between groups, even before the conquest. Given this diversity and the region's porous and plastic frontiers, experts continue to passionately debate Mesoamerican nomenclature and periodization.

The most common historical division delimits time periods according to stages of technological development and social organization and includes the Lithic period (10,000 to 3500 BCE, more or less), the Archaic period (3500 to 2500 BCE), the Formative or Preclassic (2500 BCE to 200 CE), the Classic period (200 to 900 CE), and the Postclassic period (900 CE to 1521 CE). Some scholars combine the Late Postclassic with the early years of the Spanish colony in a "Historical Period," given the persistence of many aspects of precolonial life among the Indigenous populations of New Spain, as well as the existence of resistance movements and independent Indigenous communities through the seventeenth century. Much of the information about precolonial Mesoamerica that still exists is preserved in the codices that were produced during this early colonial period (Solís 2001, 29). Despite criticism, the concept of Mesoamerica continues to inform discussions and paradigms in archaeology and ethnohistory, among other fields, and I will cautiously use the term here.

Archeoastronomy

Whether or not they could be considered "scientists" in the modern sense, Mexica, Maya, and other Mesoamerican populations did pay close attention to the night sky through observational and horizon astronomy. For example, they kept track of celestial events such as eclipses, which they were able to predict with some accuracy, comets, supernovae, and the orbits of Venus and other celestial bodies (Galindo 2009, 67). The postconquest *Codex Mendoza* includes paintings of seated observers contemplating the night sky, an activity represented by a conic symbol topped by a star-glyph connected to the observer's eye with a dotted line. Above the skywatcher is an image of the sky itself, a dark-gray semicircle "seeded and bordered" with star-glyphs, or "stellar eyes" (Ragot 2016).

In his analysis of precolonial Mayan astronomy, Gerardo Aldana y Villalobos (2021, 11) describes the activity of *ajk'ins*, "timekeepers staying up all night (in light-pollution-free environments)":

> Meteors interpreted as "arrows of the celestial deities" would certainly have been observed—sometimes in isolation, other times in "showers"—and communicated to the appropriate authorities for interpretation. . . . They noted and accounted the wanderings of the planets, following their paths past the "stars of the night" and other celestial bodies that remained "fixed" relative to each other. . . . With the *ti'huuns* [lords of the night] in their seats, with personified planets, and with the Moon's and the Sun's arrival or departure over the horizons, the celestial realm would have been viewed as populated by a dynamic community, laden with interactions—predictable and not—to be interpreted by astronomers and priests.

Nahuas and Mayans viewed the Milky Way as a road or a serpent, and both cultures occasionally used flower metaphors to refer to the night sky. According to the Postclassical *Paris Codex*, the Mayans recognized thirteen constellations that formed stellar trees, birds, and other animals. Nahua constellations included the stellar ball game (Gemini) and the scorpion (Scorpius). The Pleiades, called the Tianquiztli or Marketplace by the Nahuas, were significant, as their appearance signaled the beginning of the agricultural cycle and marked the start of the Festival of the New Fire (Milbrath 1999, 253).

Of the planets, or "wandering stars," Venus was particularly important. Jesús Galindo describes images found in the Postclassical Mayan metropolis of Mayapán that he interprets as an observational register of the transit of Venus. He describes the painting of large solar disks within which richly attired characters with blotches on their skin appear, noting that, twice a year on calendrically significant dates, the sun's rays fully illuminate the disks. He conjectures that these figures can only refer to sunspots or the passage of the interior planets. Mercury would be too small to be seen by the naked eye, but Venus is large enough to be seen without the aid of a telescope, or even a special filter, given that Mayapán's flat landscape would permit the observation of the rising or setting sun, filtered by the atmosphere itself (Galindo 2009, 69). Although not all scholars accept this interpretation, they agree that the Mayans did register Venus's synodic period of 584 days and the phases of its appearance and disappearance.

The *Dresden Codex* from northern Yucatán, which dates to the early Postclassic period, contains evidence of observations of the Moon, Venus, and Mars, as well as the register and prediction of eclipses. The particular significance that Venus had for Mesoamerican cultures stems from its dual nature as both the evening and the morning star. For Nahua populations, Venus as the morning star was identified with the deity Quetzalcóatl, the Plumed Serpent who heralds the dawn, while Venus as the evening star was personified by Quetzalcóatl's twin, the monstrous canine Xolotl, who accompanies the Night Sun as it passes through the underworld. (Notably, Xólotl is also the name that was given to the exoplanet HD 224693-b in a contest sponsored by the International Astronomical Union, or IAU, in 2020. Xólotl orbits the star Axolotl, the name of an amphibian that is native to Mexico.) Together, the morning star and the evening star were recognized as Tlahuizcalpantecuhtli, the lord in the dawn, who was sometimes linked to the rain deity Tlaloc (Montero 2022, 297). Perhaps because of its transformational nature, as well as its appearance at the start and end of the rainy season, Nahua populations often linked Venus to the maize plant, symbol of growth and fertility. In the more arid Gran Nayar, Venus was associated with deer and cactus peyote (Neurath 2004, 110). Some Mesoamerican narratives represented Venus as a dangerous hunter or warrior (always male) armed with arrows.

Mesoamerican astronomical and architectural specialists connected the movement of celestial bodies with the earthly environment, both natural and built. Like Palenque, many ritually significant architectural complexes were constructed with an intentional astronomical orientation that considered the rising or setting sun, solstices, and equinoxes, for example (Galindo 2009, 67). The orientation of the building known as the Castle in Chichén Itzá, a tourist favorite today, gave rise to the interplay of light and shadows that permits Kulkulcán, the Mayan version of Quetzalcóatl, to slither up and down the pyramid's balustrade each year on the spring equinox. The Moon's movement in the sky throughout the year was the guide for the orientation of the Temple of Ixchel in Cozumel. The alignment of the Milky Way seems to have inspired the orientation and interior artistic representations found in the Building of the Paintings in Bonampak. The ceiling of the edifice's central room is divided into four quadrants, each painted with representations of celestial objects, including a turtle accompanied by three star-glyphs that may refer to the European constellation Orion, and a band of peccaries that have been interpreted as a representation of the Pleiades (Galindo 2009, 69). As planetarium visitor Enrique mentioned, many of the buildings at Teotihuacán were astronomically oriented, also pointing toward features of the natural landscape, such as the Cerro Gordo that rises behind the famous Pyramid of the Moon (Neurath 2023, 150).

Without disregarding the importance of observational astronomy, Aldana cautions against an unduly romantic view of Mesoamerican stargazing as a disinterested protoscience that involved precolonial cosmic specialists solely dedicated to the contemplation of the night sky and the register of their observations, analogous to the activity of classical Eurasian astronomers (Aldana y Villalobos 2007, 47). Similarly, although many narratives about the history of Mexican astronomy begin with descriptions of ancient astronomical knowledge, León-Portilla (1995, 9) argues that preconquest astronomy, while vital and extraordinary, did not constitute knowledge for knowledge's sake; rather, mathematical and calendrical calculations functioned as a way to satisfy these populations' need to assure the proper ordering of time and space, "their world vision," as well as to "fulfill their survival needs." Astronomy as a science, he asserts, did not make an appearance in Mexico until the colonial period, echoing Joanna Broda's (1995, 71) assertion that pre-Hispanic astronomy "has an important historical and cultural value," but that "there was no continuity between pre-Hispanic and colonial beliefs," and that therefore "pre-Hispanic astronomy does not give us direct antecedents for contemporary astronomy in Mexico."

Ollin

In mid-December 1790, workers digging up the ground in Mexico City's main plaza came across a giant stone near the acequia that ran alongside the viceroy's palace. They were beginning to drain and pave the plaza under orders from the second Count of Revillagigedo, who was attempting to "bring new order to the chaotic and flood-prone city" (Stuart 2021, 11). When they tried to move the twenty-four-ton basalt disk, they discovered that it was not simply an inconvenient geological obstacle to progress, but an intricately carved circular monolith that had been placed facedown. Known at the time as the "Aztec Calendar Stone," the disk was set up vertically against the wall of one of the newly constructed towers of the cathedral where the public could see it. There it was analyzed and interpreted by historian and astronomer Antonio Gámez y León, as well as by foreign visitors like the German naturalist Alexander von Humboldt, who published a description of the monument in his *Voyage to the Equinoctial Regions of the New World*.[7] For decades, the Piedra del Sol stayed propped up against the cathedral wall, subject to inclement weather and "victim of the populace, who threw trash at it, and of the North American soldiers who used it as a shooting target during the days of the 1847 occupation" (López 2008, 82). In 1885, after archaeologists expressed their concern for the Sunstone's future, it was moved to the Gallery of Monoliths in the colonial Casa de Moneda, which would serve as the first National Museum. Eighty years later, it would be moved a final time "on a cement and steel platform sustained by sixteen wheels," "pulled by a 290-horsepower tractor," and "accompanied by the sound of [the song] 'Las golondrinas,'" to the National Museum of Anthropology, inaugurated in 1964 in Chapultepec Park (83). Today, it serves as the crown jewel of the Sala Mexica, itself the centerpiece of the Museum, designed to showcase a modern Mexico City constructed on the foundation of the Aztec Tenochtitlan, and a modern nation constructed on

FIGURE 4 Piedra del Sol. Photo by the author.

the foundation of a glorious Indigenous past (Rosas 2021, 135). Since 1887, when Porfirio Díaz inaugurated the Gallery of the Monoliths, the Piedra del Sol has been "symbolically identified with the number one in the museum catalog" (Achim 2017, 251).

The Piedra del Sol's origins are unclear, but it was probably dragged by hundreds of men to Tenochtitlan from several miles away in the late fifteenth or early sixteenth century. The stone was erected in the Templo Mayor, or possibly a smaller temple, although it was removed from its sanctuary after the conquest and left for decades faceup, on the ground near the Viceregal Palace. But in the mid-sixteenth century, the idolatry-obsessed archbishop don Alonso de Montúfar "saw the evil and murders that happened there, as well as other things he suspected" and ordered the monument to be buried facedown, "so that the memory of ancient sacrifice would be lost" (Duran, in López 2008, 80). The stone stayed buried for two hundred years.

The Sunstone's basic geometrical form is the quincunx, the quintessential cosmological symbol in preconquest Mesoamerica, repeated in the expressive culture of the populations that inhabited what is now Mexico over thousands of years before the coming of the Europeans in the sixteenth century. Carved into buildings and monuments from the Gulf Coast, the Yucatán Peninsula, Oaxaca, and Mexico's central valleys, in the material culture of the Olmecs, Mayans, Zapotecs, and Mexicas, the symbol appears as a simple geometrical decoration, as a design element in the garments of deities, in the form of four-petaled flowers and intertwining serpents, and as a model for the human body. The least complex version of the quincunx, which represented the four directions and the center of the cosmos, was the image of five discrete points. It also appeared as two bands that crossed in the middle, as a square with a circle in the center, and as a square with a center circle and circles in each corner. The quincunx serves as the basis for iconographically complex unions of the glyphs that refer to the movement of space and time.

Most scholars have interpreted the Piedra del Sol's center as the face of Tonatiuh, the fifth Sun, known as the Sun of Movement. Two circles representing the creator couple Omecihuatl-Ometecuhtli (also known as Tonacatecuhtli-Tonacacíhuatl) mark the deity's forehead, his ears are pierced by eagle feathers, and his tongue is carved into the form of a sacrificial jade knife. His clawed hands emerge from the left and right of his face, each holding a human heart.

A series of concentric circles surround this image. The first circle is the *nahui ollin*, the symbol of cosmic movement, considered by scholars to represent a stylized ball-court (Neurath 2023, 44). The symbol frames a group of icons representing the creation and destruction of the first four suns:

> Six hundred years after the birth of the four brother gods, the sons of Tonacatecli [Tonacatecuhtli], they got together and said it would be good to order what had to be done, and they all charged Quezalcoatl [Quetzalcóatl] and Uchilobi [Huitzilopochtli] with the ordering of things. (García Icazbalceta 1891, 229)

The anonymous author of the *Historia de los Mexicanos por sus pinturas* goes on to explain how Quetzalcóatl and Huitzilopochtli created fire and then made a semi-sun, which, because it was not whole, did not shine brightly. The gods proceeded to make a man and a woman, to whom they gave seeds of maize. They made the god and goddess of the underworld, then the skies, up to the thirteenth level, and they created water. In the water, they transformed the giant lizard Cipactli into the earth, Tlaltícpac. Cipactli was divided into two halves, and each half into nine pieces. The top nine were the layers of the sky, and the bottom nine were the layers of the earth and the underworld. Then the four gods together created the water god and goddess, who have a house of four rooms, with a patio in the center that holds four great containers of water from which pour four distinct kinds of rain with distinct consequences for humans and their crops.

All this occurred in an undifferentiated time because the count of years had not begun. The gods saw that the half sun was not bright enough, so Tezcatlipoca became the sun, and they created the solar calendar, dividing it into eighteen periods (like the eighteen pieces of Cipactli's body) of twenty days, and time began to be counted. This first sun, the Jaguar Sun ruled by Tezcatlipoca, came to an end when the sky fell to the earth, the sun stopped its movement, the day became night, and humans (who may have been giants) were devoured by Tezcatlipoca in the form of a jaguar after the god had been knocked out of the sky by his brother Quetzalcóatl, who then became the second sun. This period was called the Wind Sun, which ended after the world was destroyed by hurricanes and humans were turned into monkeys after climbing trees to try and save themselves. Tezcatlipoca expelled Quetzalcóatl from the sky, and the rain god Tlaloc became the third sun. The Rain Sun ended after a shower of fire sent by Quetzalcóatl. Volcanoes and lava destroyed the earth, and the gods transformed the humans into birds. The rain god's wife Chalchiuhtlicue became the fourth sun, the Sun of Water, which ended after a world flood. The gods turned the only people left into fish. After the fourth sun, the gods made four paths into the center of the earth, creating four men to help them. The four gods became trees,

> . . . and men and trees and gods raised up the sky with the stars as a roof, and for having done this, Tonacatecli their father made them lords of

the sky and the stars; and because the sky had been raised, Tezcatlipoca and Quetzalcóatl made the path that appears in the sky, where they met, and are afterward there in it and with their seat in it. (García Icazbalceta 1891, 234)

The gods created the Sun of Movement, the center of the Piedra del Sol, when they came together in Teotihuacán to lament the lack of light in the world. They decided to make a new sun, which would feed on human hearts and drink human blood. Two people offered to sacrifice themselves, the rich, handsome Tecuzitecatl and the poor, diseased Nanahuatzin. Nanahuatzin made the proper offerings, although they were humble, and immolated himself eagerly to become the sun. Tecuzitecatl made rich offerings but had to be coerced into throwing himself in the fire, so he became the less luminous Moon. The Fifth Sun, in which the inhabitants of Mesoamerica were still living at the time of the conquest, will eventually be destroyed by earthquakes.

The second circle of the Piedra del Sol contains the glyphs of the count of days, while the third ring is made up of repeated quincunxes representing the cosmos and the heat of the sun, anchored by a series of triangles that symbolize the locations of the four regions of the cosmos, separated by solar rays. A circle of sacrificial elements that underscore the importance of the circulation of blood for maintaining the heat of the sun surrounds the ring.

The next circle is composed of two fire serpents whose tongues are fused at the bottom of the sculpture. From their open mouths emerge two deities, whose identities are debated, possibly Tonatiuh and Xiuhtecuhtli, and whose headdresses include symbols of the Pleiades. In Mexica cosmology, the two serpents, which personified the Milky Way, carried the sun through the sky from the east where the serpents' tails meet at the top of the circle, as it passes through the daytime sky into the underworld where, as the Night Sun, it illuminates the sky of the dead.

Surrounding the serpents are two more rings symbolizing the nocturnal and diurnal skies. The rim of the disk is decorated with "star symbols representing the night sky, the realm of Yohualtecuhtli, the deity of the nocturnal sun" (Navarrete 2019, 753). The icons only partially circle the stone, as its carvers had been unable to separate the sculpture from the surrounding rock because of a fracture. The stone was probably painted in fiery yellow and red tones, highlighting its solar imagery (Solís 2000, 39).

Federico Navarrete (2019, 753) argues that "the iconography of the monument was so exhaustive because its function was to assemble, organize and control all these different ontological, spatial and temporal aspects" that characterized the Mexica cosmos. The flatness of the stone, Navarrete continues, represents the horizontal region of the universe: the *tlaltícpac*, or the surface of the earth. The "top" would have been oriented toward the east and the "bottom" to the west. Although the Piedra del Sol is displayed vertically at the National Museum of Anthropology, most scholars believe that it may originally have been mounted horizontally and used as a base on which the ritual sacrifices that assured the passage of time were performed. Thus, the Piedra del Sol is not only a symbolic representation of the cosmos but a material agent that plays an active role in its regeneration. The Mexica believed that the eventual destruction of this world was inevitable, but that humans had the obligation to attempt to keep the sun alive as long as possible through sacrifice, providing the sun with food in the form of the blood of women who died in childbirth, warriors killed in battle, and captured prisoners. Ritual sacrifice was a form of reciprocity: the gods nurture and guide humans, and therefore humans are responsible for making offerings to sustain the gods.

Cosmic Chaos

The Piedra del Sol was not the only monument unearthed during the undertaking of public works in Mexico City's main plaza in the late eighteenth century. A second sculpture was also examined by Alexander von Humboldt during his Mexican expeditions and is also exhibited today in the Sala Mexica of the National Museum of Anthropology. Unlike the Piedra del Sol, this monolith is not a symmetrical simulacrum of the aesthetically ordered cosmos. The giant figure had originally been connected to Teoyamiqui, a goddess of death, and the *tzitzimimeh*, female deities associated with the stars who would descend to earth to devour humans in the event of an eclipse or the destruction of the sun. Today, scholars usually identify her as Coatlicue, an earth deity considered to be the mother of the gods Huitzilopochtli, the Mexicas' divine patron,

Coyolxauhqui, the moon goddess, and their brothers, the four hundred southern stars. She was the wife of Mixcóatl, a deity associated with war, hunting, and the Milky Way, and is sometimes referred to as Tonantzin, "our mother." For Antonio León y Gama, Coatlicue was a "horrible simulacrum," for Humboldt, a "monstrous idol," for George C. Vaillant, a "dynamic concentration of all the horror in the universe," and for Salvador Toscano "the masterwork of American sculpture" (García Ramírez 2021). The eight-foot-high statue depicts a decapitated and partially dismembered woman with two joined snakes instead of a head, clawed hands and feet, pendulous breasts, a necklace of human hearts and hands, a skirt made of interwoven snakes, and a skull at her waist. On the base, the anonymous artist or artists carved the figure of Tlaltecuhtli, the Lord of the Earth.

Navarrete (2019, 759) contrasts the statue of Coatlicue with the Piedra del Sol, which "can be easily recognized as [a cosmogram] that aim[s] to represent and maintain order and regularity in the world and its chronotopes." The female figure, however, is deliberately ambiguous, joining different temporal and spatial orders:

> As an earth goddess, she belonged to the underworld, and to the distant past, before the creation of the current solar chronotope. She embodied the chaos that needed to be destroyed so that the solar order could emerge, and the sacrifices that enabled the creation of the space-time inhabited by the current sun, and by the Mexica. The monument would reproduce and reenact this defeat, to celebrate and confirm the supremacy of the solar chronotope. In contrast, as a *tzitzímitl*, the monument belonged to the heavens, from which these terrible beings would descend at the end of the current era. It established a link with the future apocalyptic time after the solar chronotope and the Imperial chronotope would collapse. Hence, it anticipated the inevitable victory by the forces of destruction that lurked beyond the cosmos that was so painstakingly maintained by the Mexica. (760)

Unlike the Piedra del Sol, Coatlicue did not remain on display after her discovery in 1790: she had been taken to the National Autonomous University for conservation and study but was soon reburied in the university patio to discourage superstition and idolatry, as the Indigenous

FIGURE 5 Coatlicue. Photo by Melissa Biggs.

inhabitants of Mexico City had begun to take her offerings of flowers and candles. She was dug up for Humboldt in 1803, then reburied, then dug up again so that the English entrepreneur William Bullock could make a mold for the Egyptian Hall in the new Piccadilly Circus, then buried once more. She was disinterred a final time after Mexico gained its independence from Spain and exhibited in the National Museum (Achim 2017, 48), where she has become one of the museum's most popular attractions and the recipient of offerings made by contemporary Indigenous groups (Biggs 2011, 174). Perhaps this ambivalence was due to what Octavio Paz (1995, 77) saw in Coatlicue: the maximum expression of radical American alterity.

Spacetime

James Maffie (2013, 15–16) argues for the indissolubility of time and space, or "time-place" in precolonial Mesoamerica. Time-place is how *teotl*, cosmic power or energy, moves. It is processual, relational, and imbricated in three patterns of "motion-change": ollin, as we saw in our discussion of the Piedra del Sol, was cyclical change, "exemplified by bouncing balls, pulsating hearts, respiring chests, earthquakes, labor contractions, and the daily movement of the Fifth Sun." *Malinalli*, a word referring to a plant with a twisted shape used to make fiber, but also a deity associated with witchcraft, possibly a daughter of Coatlicue as well as a day-sign in the Mesoamerican calendar, was spiral motion-change, "exemplified by spinning fiber into thread, cooking and digesting food, blowing life into things, drilling fire, burning incense, and ritual music, speech and song." Ollin refers to change within cycles, while malinalli is characterized by change across cycles, conditions, or spaces. The third pattern of motion-change was *nepantla*, "exemplified by mixing and shaking things together, weaving (interlacing), and sexual commingling." Nepantla joins ollin and malinalli so that teotl can continue to unfold and transform. If the woven/weaver/weaving cosmos is teotl, argues Maffie, nepantla is the mechanism, ollin is the horizontal weft, and malinalli is the vertical warp.

Maffie considers Mesoamerican calendars to be the warp patterns of their age. Most groups counted time using a combination of two calendars. The first was a 365-day solar calendar, called the *xiuhpohualli* by Nahuas and *haab* by the Maya, which was divided into eighteen periods of twenty days each, with five extra days. For the Mexica, these "unnamed" days marked the welcoming of the new year with the ritual of the New Fire in Iztapalapa. The Maya grouped these days into an extra month, called *Wayeb'*, during which the population was expected to abstain from bathing, eating, and engaging in sexual relations. The second was a 260-day calendar that combined the twenty "month" signs with another series of thirteen numbers to form a date, making 260 possible combinations. The solar xiuhpohualli was based on astronomical observation, but the use of the 260-day *tonalpohualli* (*tzolk'in* for the Maya) was closer to the practice of astrology, although without reference to the stars (Díaz Álvarez 2018, 85). Each date (and those born on that date) received the influence of two deities and one direction. Scholars have

proposed various explanations for the significance of 260 days; one of
the most popular is the period of human pregnancy. The two calendars
began on the same date but were desynchronized after 260 days, linking
up again after fifty-two solar years or seventy-three tonalpohualli cycles.
The Olmecs, Mayas, and Zapotecs had a third calendar, called the "Long
Count," a linear timeline that began in August 3114 BCE that Aldana y
Villalobos (2007, xxviii) refers to as "an odometer of days." In addition
to its permitting the prediction of cosmic events, like the periodicity of
Venus registered in the *Dresden Codex*, Alonso Zamora proposes that
the Long Count had ritual significance, providing a temporal framework
in which Mayan leaders could "time travel" to the past and the future to
participate in rituals and negotiate with the gods (in Neurath 2023, 212).[8]
However, the uses of this calendar are still not completely understood,
and it had fallen out of use by the Late Postclassic period.[9]

Colonial chroniclers often imagined Mesoamerican time as a circle or
a wheel; one version of the clockwork cog image of the relationship between the 365- and the 260-day calendars presented in the planetarium
film *Mayan Archeoastronomy* was painted in the Yucatan Maya *Chilam
Balam*, while another can be found in the central Mexican *Codex Mexicanus*.[10] However, some scholars have argued that these circular representations were derived from spherical Aristotelian-influenced models of
celestial motion (Aveni 2012, 11). Despite the existence of round objects
like the Piedra del Sol, most images of time in precolonial Mesoamerica
take the form of a square or a rectangle, with time represented in the
corners as the points of solstitial sunrise and sunset. Dates were also arranged in precolonial almanacs linearly, in a zigzag, cruciform, or spiral
pattern, in the contours of animal shapes, or even randomly (73). The
image of the milpa, with four trees marking the corners of the cultivated
field, or a house, with four pillars sustaining the roof, also contains the
notion of time, located in the four corners. This connection continues to
inform contemporary Indigenous cosmology (73).

Astorga compares the notion of *tlacauhtli*, "space" in general, with
"*altépetl*," which designated a particular settlement and its political organization, and "*tlalli*," the earth, both Earth and "the earth where I am
standing," a notion related to property and economic productivity. The
three concepts are distinct but related through the notion of ordered
movement, from the cosmological to the political (Astorga 2014, 57).

Spatially, scholars have tended to represent Mesoamerican cosmograms with a straightforward model of four directions and at least three superimposed levels. For the Mexica, the surface of the earth, Tlaltícpac, was a great disk surrounded by water. They called the aquatic ring Cemanáhuac, the origin of the modern use of the name "Valley of Anáhuac" for central Mexico. The water that circled the earth was called *teoatl*, divine water, or *ihuica-atl*, celestial water, as it joined the heavens at its horizon. In the words of Miguel León-Portilla (2017, 169), the four geographical extensions of Cem-anáhuac are surrounded by "hives of symbols":

> The Nahuas described them facing the west and contemplating the passage of the sun: that way, where it sets, is its house, the country of the color red; then, to the left of the sun's path, is the south, the direction of the color blue; in front of the region of the sun's house is the direction of light, fertility, and life, symbolized by the color white; finally, to the right of the sun's route extends the black quadrant of the universe, the direction of the land of the dead.

The most well-known image of the Mesoamerican vertical cosmos, and the one learned by Mexican schoolchildren, describes the universe as having nine or thirteen upper levels and nine lower levels, an idea based on the Mexica *Lápida de los Cielos*, or "Stone of the Skies," and a scene from the early colonial *Codex Vaticanus A*, among other sources. In this model, the first level of the sky is called Ilhuícatl Meztli, the space traversed by the Moon and that which upholds the clouds. It is also the home of Tlaloc and other beings associated with water. The second level is Citlalco, the home of the four hundred (that is to say, innumerable) northern and southern stars. Through the third level, Ilhuícatl Tonatiuh, the sun traces his daily path. Ilhuícatl Huitzlan, the fourth, is the home of Venus. The comets, or smoking stars (*citlalin popoca*), inhabit the fifth sky, while in the sixth and seventh can only be seen as the colors green and blue. The eighth is the home of storms, while the ninth through the eleventh levels, the Teteocan, house the gods. The upper two skies, according to Miguel León-Portilla (2017, 162), constitute the metaphysical Omeyocan, the mansion of duality, where Ometéotl, the primordial deity reigns.

The lower nine levels are the regions of the underworld through which the dead, or the "fleshless," must pass through a series of trials for four years before they can finally rest in Mictlán, the lowest level (León-Portilla 2017, 169). There were, however, other possible cosmographical destinies for the dead. Tonatiuh Ilhuícatl, the house of the sun, was the destination for those warriors who died in battle and would be allowed to accompany the sun on its journey from dawn until midday when they would be relieved by the other class of dead who lived in this level of the sky: women who had died in childbirth. To Tlalocan, the Paradise of Tlaloc, traveled those who had died by drowning, had been struck by lightning, or had been afflicted with any of the diseases associated with water, such as leprosy, gout, or edema. Small children who had died before being weaned went to Chichihualcuauhco, the nursing tree, where they awaited their turn to be reborn (Salas and Talavera 2010). Tlaltícpac is often represented as simultaneously the first of the vertical upper layers and the first of the vertical lower layers, as well as the center of the horizontal cosmos; the space inhabited by humans is thus implicated in the nonhuman (or pre- or posthuman) cosmos.

However, Mesoamerican conceptions of the cosmos were much more complicated than static models can express. The content of each level varies from source to source, and stratified layers are not always present. Díaz Álvarez argues, for example, that the number nine appears much more commonly than the number thirteen in reference to the levels of the sky, and that the location of the pair of creator gods often varies. Linguistic and graphic analysis also shows, she adds, that the division of the sky into nine parts may have been conceived as a series of crossings or meetings of roads rather than a series of superimposed "floors" (in Mikulska 2020, 277). The "underworld" was not necessarily fixed in place; it may have also been linked to the night sky "above us" and the night sun, which, with the starry sky, passes "below us" during the day (Mikulska 2008, 164), to the interior of the earth represented by pyramids, caves, and cenotes and, as we have seen, to the horizontal direction "north" (or south, for some populations).

Nielsen and Sellner Reunert (2020, 45) remain convinced that, even if the existence of a basic three-level structure is evident, the quadripartite horizontal model is the essential characteristic of the Mesoamerican cosmos. Analyzing a wide variety of painted and engraved cosmic representations that predated the conquest, Díaz Álvarez concludes that the divisions and unions of "upper" space were not necessarily organized as stratified layers, but a complex segmentation of distinct spaces that can be considered "modular" in the same way as the altépetl, or political entity, the *calpulli*, or local community, and the *milpan*, or cultivated parcel. The mirroring of the celestial/infernal and the earthly realms had local and regional variations, according to particular sociopolitical and historical geographies, "in contrast to a universal (theological or scientific) cosmological worldview" (Díaz Álvarez 2020a, 115).

The ontological status of the cosmic realms was also flexible; the sky was seen at times as "a container—as a place full of life, afterlife, and movement, where the lords, gods, or ancestors ruled and had an impact on earth when they decided to interfere in human history" (Díaz Álvarez 2020a, 118). In other contexts, the sky was a body composed of organic material: half the Cipactli, the giant caiman from whom the earth was formed, or, in another narrative, of the god Tlalteotl, while the stars were eyes that irradiated light. "The body is a microcosmos; the world is a macrobody" (Neurath 2021, 164). The sky could also be a frame or a threshold, a series of borders separating levels or regions. However, these boundaries were porous, permitting transit and communication between the distinct places in cosmic space through such devices as portals, often found in caves, temples, or cenotes, or even "cosmic sphincters" that combine the characteristics of the sky as body and threshold (Díaz Álvarez 2020a, 119).

Díaz Álvarez and other scholars have questioned whether the model of thirteen heavens and nine hells is, at least in part, a colonial imposition rather than an accurate representation of precolonial visions of the cosmos. As she points out, the model of the cosmos based on a series of superimposed layers is not unlike the structure of the Christian universe envisioned by Dante, and was logically embraced by Spanish friars during the colonial period, as its "pedagogical efficiency and its similarity with classic Eurasian cosmographies eased the model's acceptance and its institutionalization as the fundamental cosmological structure shared by all Mesoamericans before and after the conquest" (Díaz Álvarez 2020b, 3). Indeed, Spanish missionaries used the words Ihuicatl and Mictlán to translate the Christian "heaven" and "hell," while Tlaltícpac continued to be used to refer to the terrestrial plane. In the colonial context, the complexity of Ilhuícatl and Mictlán, references to those worlds that are beyond, *más allá*, "there" (the night or daytime sky, dreams, the remote past) rather than "here" (the present, waking life, the earth), was lost in translation.

Cosmic Diplomacy

Precolonial Mesoamerican cosmograms are complex representations of the dynamic universe—which might more properly be considered a multiverse, or at least something other than a "monocosmos" (Martínez Ramírez and Neurath 2021, 8)—inhabited by humans and nonhumans. However, graphic depictions and religious doctrine were not its most important components. Unlike "Christian perennial metaphysics," precolonial Mesoamerican cosmology was "generative" (Díaz Álvarez 2020a, 115), as the cosmos required human intercession for its continued existence. Sacrifice was fundamental in this sense, a cosmic praxis that attempted to ensure the circulation of vital energy, teotl, between the visible and invisible, human and nonhuman worlds.

Maffie (2003, 76) writes of "a participatory universe" generated through "a relationship of compelling mutuality" between humans and the cosmos. He characterizes precolonial Nahua philosophy or *tlamatiliztli* as the interrelation between four "practical abilities."

> First, it consists of the practical ability to conduct one's affairs in such a way as to attain some measure of equilibrium and purity—and hence some measure of well-being—in one's personal, domestic, social, and natural environment. Secondly, it consists of the practical ability to conduct one's life in such a way as to creatively participate in, reinforce, adapt, and extend into the future the way of life inherited from one's predecessors. Thirdly, *tlamatiliztli* consists of the practical ability to conduct one's life in such a way as to participate in the regeneration-cum-renewal of the universe. Finally, it consists of the practical "know-how" involved in performing ritual activities which: genuinely present *teotl*, authentically embody *teotl*, preserve existing balance and purity, create new balance and purity, and participate alongside *teotl* in the regeneration of the universe. (76–77)

According to Maffie, precolonial Nahuas would consider much of Newtonian science to be "unrooted in *teotl*." By contrast, Nahua epistemology is based on knowledge acquired through personal acquaintance. "It does not pursue goals such as truth for its own sake, accurate representation, empirical adequacy, or manipulation and control; nor is it motivated by questions such as 'What is the (semantic) truth about nature?' or 'How can we master and bend the course of nature to our will?'" (Maffie 2003, 78). Ritual action was not a way to dominate the forces of the cosmos, but rather to "cooperate with the universe in order to renew the universe" (78).

Through cosmic diplomacy (Kohn 2020; Battaglia 2014), humans had to constantly negotiate with nonhuman agents through sacrifice, ritual speaking, and creative production, in a kind of existential uncertainty principle that allowed for flexibility in the face of extreme adversity. As Neurath argues (2023, 65), the gods were close to humans, belonging, in a certain sense, "to an expanded human community." They may have been ancestors or relatives. Gods could become humans, and humans could become gods. As *ixiptla*, the gods were present in the world and were produced through human ritual action (54). However, they were also located "elsewhere" and considered dangerous and mysterious "others" that were very difficult to control.

> In your relations with these deities, the best you can do as a human is maintain the status quo. . . . Perhaps we can say that the goals of Mesoamerican religious practice are modest: trying to survive a little longer, trying not to die. They don't have anything to do with achieving eternal life, liberation from all evil, or salvation. (53)

Cosmic Portents

> Around midnight . . . one of the guards saw a smoke rise toward the east, getting denser, it was so white that it shone and gave off such light that it seemed like midday . . . and it grew, joining together the sky and the earth, so that it seemed to be walking like a great white giant. He quickly called his companions . . . saying to them, "It is not your duty to sleep but to watch. Get up and you will see what this is that came out of the east,

almost stuck to the sky, a smoke so white, like a dense white cloud." And all
who watched from the temple saw it and were attentive until dawn came,
and then it started to disappear little by little until it vanished into nothing.
(Alvarado Tezozómoc 2021, 576)

When told of this portent a few years before the Spanish conquest, the
tlatoani Moctezuma II was greatly perturbed. Nezahualpilli, the poet-
king of Texcoco, was called to explain it to him. "How are you ignorant
of this?" he asked Moctezuma. "How have you not been told by those
who guard the city and watch the sky and the stars?" The portent, he
said, foretold the end of this world. And it would not be the last sign that
disaster was coming; it was among other "singular events that announced
the arrival of the Spaniards" (577). The comet would be the fourth of
eight portents, or *tetzáhuitl*, each considered "an anomaly, a rupture of
harmony, a little bit of chaos slipping into ordered reality" (Burkhart
1989, 64). In addition to the comet, the portents that announced the conquest included lightning striking the temple of Xiuhtecuhtli and other
unexpected fires, flooding, the cries of the snake-woman Cihuacóatl in
the night, the appearance of two-headed monsters, and the capture of a
crane with a mirror on its head. In this mirror, Moctezuma II saw "the sky
and the stars, and especially the *Mastelejos* [the "fire drill" constellation
in Taurus] that appear near the *Cabrillas* [the Pleiades]," and then "a great
crowd of people who were armed, riding horses" (although the Nahuatl
text uses the word for "deer"). The apparition of stars in the mirror, particularly the Mastelejos, "announces a change of cosmic eras" (Olivier
2019, 34).[11] Indeed, some authors argue that the spiral historical logic of
birth and destruction of successive solar eras dominated in turn by Quetzalcóatl and Tezcatlipoca was the basis for the belief that Cortés could be
a personification of Quetzalcóatl, returning to announce a new sun (38).

The friar Juan de Torquemada, like many of his evangelizing contemporaries, was quick to accept Indigenous narratives of otherworldly signs.
In 1589, he wrote: "In difficult cases and complicated affairs that for God's
just judgments happen in the world, there are often signs and prodigies
that prognosticate these events before they happen, especially at the end
and desolation of some kingdom" (in Pastrana 2004, 16). As various authors have noted, the colonists' own medieval acceptance of cosmic portents made it easy for them to believe that comets and other catastrophic

signs were divine justification for their colonizing and evangelizing missions in "the New World" (17; Olivier 2019, 32). Spanish chroniclers of the conquest also reference miraculous visions, such as the Virgin Mary, "beautiful and resplendent as the sun," whose appearance frightened off the Mexica defenders during the battle of the "Noche Triste," when the Spaniards were forced to flee Tenochtitlan (46).[12]

As we saw in the case of the cosmic portents that presaged the fall of the Aztec empire, both the Spanish and the Indigenous populations of New Spain believed in some form of astrology. Medieval European almanacs, calendars, and *repertorios* combined astronomy, geography, chronology, and natural philosophy. Many of these books and pamphlets prognosticated the future on the basis of witnessed or predicted astrological influences on terrestrial processes. Astrology, considered to be both a science and an art, "was one of the great bodies of knowledge, or sciences, of the late medieval and early modern periods. Individuals from all social strata and levels of education accepted it as a way of knowing" (O'Hara 2018, 47). Professional cosmologists like Enrico Martínez (Heinrich Martin), who held the first chair in astrology in New Spain, combined their knowledge of the general science of the stars, "natural astrology," with the "interpretation of astral events on earth," known as "judicial astrology," which is the term Spanish chroniclers would use to refer to Indigenous divination practices (4; Mundy 2023, 396).

In some cases, much like the Dantesque structure of the Mesoamerican cosmos captured in early colonial paintings, paintings of comets blazing across the sky on the eve of the conquest are clearly inspired by medieval European sources. Some authors have taken the European influence on stories of comets and other portents as evidence that these were completely Spanish inventions rather than Indigenous interpretations of the apocalyptic conquest. Felipe Fernández-Armesto (1992, 288), for example, writes:

> Considered from one point of view, the conquest of Mexico was, it seems to me, a clash of equally aggressive, equally dynamic, equally self-confident warrior-societies, the outcome of which was nicely balanced. By blotting out the dazzle of comets and celestial signs, which still blind some historians, we can put the Aztecs in a clearer light, their aggression uninhibited by omens, their strength undrained by fatalism.

The postcolonial critique of after-the-fact descriptions of ominous cosmic portents announcing the Apocalypse is an important corrective to a simplistic acceptance of these descriptions, often produced by social actors heavily invested in the Christian colonial order. However, a more nuanced analysis is warranted, particularly given the cosmic portents that announce the fall of political leaders that are found in established precolonial sources (Aimi 2009, 69–147). Ultimately, the authors of after-the-fact testimonies of ominous portents announcing the end of their world lamented their loss by deploying a discursive structure that was common to both precolonial Mesoamerica and medieval Europe.

In any event, a more restrained, "rational" astrology was emerging as early as the sixteenth century, as evidenced by a polemic between Sigüenza (who, after studying to be a Jesuit, held the position of royal cosmographer and professor of astrology and mathematics) and fellow Jesuit Eusebio Kino about the true nature of comets. Like preconquest Indigenous astronomers, Kino argued that a comet that had appeared in the skies over New Spain in 1681 was a sign that something terrible was about to happen. But in his text *A Philosophical Manifesto Against Comets, Freed from the Power They Have over the Timid*, Sigüenza argued that comets had little power over what happens on Earth. He appealed to rational thought that, he noted acerbically, was not limited to Europeans, although many thinkers of the time believed that the hot climate of the Spanish colonies caused even Europeans born and raised in the Americas to "adopt the indolence and vices attributed to the Indians" (Fernández 2004, 67).

Colonial Spacetime

As we have seen, Spanish missionaries appropriated Indigenous terms as a means of translating and transmitting a model of the Christian spatial cosmos. Order became allied with "good" and chaos with "evil"; the sun and *ilhuicatl*, now reduced to a singular "sky," were allied with Jesus Christ, and everything associated with the night or the underworld were linked to the devil (Burkhart 1989, 41). Teotl was glossed as divinity. The multilayered universe was assimilated to late-medieval-hierarchical cosmograms, like Diego de Valadés's image of the Great Chain of Being.

And the idea of a four-part cosmos had an echo in a different universal cartography. In the seventeenth century, the minor Indigenous noble Chimalpahin wrote, "All of the world's lands discovered up to now are divided into four parts. So that those who see this text will know, the first part is Europe; the second, Asia, the third, Africa, and the fourth, the New World." The capital of this world was Rome, and its "universal lord" was the king of Spain (Gruzinski 2010, 91).

In general, the Spanish maintained the precolonial organization of terrestrial space and the architectural logic of power, erecting their own religious and political edifices on top of the ruins of older sites. The new regime "occupied places already saturated with meanings, and, like a parasite, fed on these already established meanings" (Mundy 2023, 369). The cathedral was built on top of the Templo Mayor, although its orientation was changed from east–west to north–south, following European cosmic conventions (374). Conquistadors lived in the palaces of the Mexica elite, and new constructions repurposed the stones of the buildings they replaced. Tenochtitlan, the center of the Mexica cosmos and the capital of the Mexica empire, became Mexico City (Mexico-Tenochtitlan, for a long time), the capital of the Spanish American empire and a geographically peripheral but cosmologically vital node in the production of globalized modernity (Quijano 2000, 533).

Although the spatial order could be adapted to the new context with relative ease, the temporal order presented a problem. The cyclical liturgical calendar notwithstanding, Christian time is fundamentally linear, with a beginning and an end. It also admits the notion of a transcendental eternity outside the flow of time in the world. In contrast to European time, colonial Mesoamerican time was a spiral, with both cyclical and linear aspects: "history does not repeat itself, but the same patterns reappear; what went before is not replaced but incorporated into new but familiar sequences" (Burkhart 1989, 72). Barbara Mundy argues that, rather than the imposition of new spatial practices, it was the institution of a new temporal order that had a greater impact on Mesoamerican life.

Indigenous elites attempted to negotiate with the emerging order, engaging in practices consistent with precolonial values of cosmic diplomacy. Mundy points out that the unknown authors of chronicles like the colonial *Codex Aubin* continued to frame their manuscripts, literally and figuratively, with precolonial calendrical glyphs. This account, written

in Nahuatl on European paper with painted illustrations, registers over
four hundred years of historical events in the Mexica world, from the
mythical migration from Aztlán through the commemorations held in
Mexico City to celebrate Habsburg rule in the sixteenth century, thus
setting "Habsburg events into a much longer span of trans-imperial history, one that the Nahua elite were quite familiar with" (Mundy 2023,
383). The authors combine the precolonial calendar glyphs with the Latin
numbers of the European calendar, attempting to correlate both time
scales. Significantly, European years are noted in the margins of the pages
dedicated to the events preceding and immediately after the conquest.
But as the chronicle moves forward in time, the dates of the Julian, and
eventually the Gregorian, calendar become more and more prominent.
It is worth noting that a number of Indigenous books containing precolonial calendrical counts were burned by the zealous bishop Diego de
Landa in Yucatán in 1562 (399).

Habsburg time was not easily assimilated into the rhythms of everyday
life, especially in the early years after the conquest. A seven-day week was
substituted for the five-day market schedule, European civic commemorations disordered the annual cycle, and the Christian festival calendar
revolving around the lives of the saints was imposed on the Indigenous
twenty-day ritual calendar (Mundy 2023, 397). The elegant organization
of months into twenty days was altered by new associations with the
irregular phases of the Moon (Aveni 2012, 33). This challenge to cosmic
order was devastating, but perhaps not entirely unexpected. For precolonial Mesoamericans, "order was temporary and incomplete, with chaotic
forces dwelling at its interstices and peripheries. Order and chaos, structure and antistructure, were subsumed within a larger pattern. Life came
from death, creation from destruction" (Burkhart 1989, 37).

The process of colonization caused a biological genocide and destroyed long-standing political, economic, and social structures. However, it would be incorrect to assert that the process implied only active
oppression on the part of the colonizers and passive acceptance on the
part of the colonized, or that all traces of precolonial beliefs and practices were erased. Rather, the three-hundred-year period between the
Spanish Conquest and the War for Independence was a time of complex
and often creative cultural negotiations—another example of cosmic

diplomacy—that reaffirmed the Mesoamerican belief in a processual cosmos in constant transformation. The dialogical exchanges between Spain and Mesoamerica, while often violent and always unequal, took many forms. Some structures had parallels in both worlds and could be translated with minimal friction. Other structures were radically opposed and were ultimately untranslatable. In these cases, the outward forms might change, but fundamental concepts would often remain.

Nepantla

During the colonial period, the Indigenous inhabitants of what had become New Spain continued to participate in practices that reflected their adherence to an animate, relational cosmos. By necessity, however, they learned to combine certain elements that had characterized precolonial Mesoamerica with elements of their new reality, often in creative ways that allowed what the Spaniards considered to be "idolatrous" beliefs to escape censure. In 1581, when Fray Diego Durán questioned a native informant about a fiesta he had organized that included the celebration of non-Christian rituals, the anonymous man replied, "Father, do not be afraid, because you see, we're still nepantla." Offended, Fray Durán insisted that he explain himself. Exactly what were they in the "middle" of?

> He told me, that as they were not yet well-rooted in the faith, I should not be taken aback, that they were neutral, that they sometimes followed one law, sometimes another, by which they meant that they believed in God, and that they also followed their ancient ways and the ways of the devil, and that's what he meant in his abominable excuse that they were still in the middle and were neutral. (In León-Portilla 1976, 80)

Centuries later, Miguel León-Portilla (1976, 80) would coin the term *nepantlismo*, which he conceptualized as a state of existing in the middle (the milieu!), the old ways ripped away, but the new ways not yet assimilated. For León-Portilla, being in nepantla was an open wound, a way of existing in the world without a sense of direction, neither here nor there. Nepantla, for León-Portilla (1962), was "the keyword for the tragedy of a people," and he extended the use of the concept to describe the experience of other collective actors, such as twentieth-century Mexican Americans, "*pochos*," who faced discrimination for being neither one thing nor another. The term was later reappropriated and resignified by authors such as Gloria Anzaldúa (1987), who used it to describe the creative potential of practices shaped by movement across and within borders.

Nepantla was also one of the great cosmic ideas of the preconquest Nahuas. As we saw, Maffie (2013, 15) considered it to be the third form of energy, tying cyclical ollin and spiral malinalli together. To be nepantla did not imply neutrality, as Durán understood it, but an undefined timespace, in the center of things, in between (or outside of) stable categories or ontologies. Following the linguistic analysis of Karttunen and Lockhart (1976), Troncoso (2011, 392) describes it as "where reciprocity is abundant" or "where mutualism is developed," and argues that it is that reciprocity that situates nepantla as a midst or a mediator between actors and elements, turning it into the "center" or "the point where elements unite." The elite Indigenous authors of the *Codex Aubin* were attempting to operate from a place of nepantla in this sense. Of course, as Troncoso affirms, when applied to intercultural dialogue, this "union" was not necessarily harmonious. Nepantla was not a goal, but a process of constant negotiation that may be a more useful category than "mestizaje" or "syncretism" as an approach to the kinds of mutual but unequal appropriations that have characterized Middle America since the conquest (395).

In many cases, nepantla meant that European forms were adopted without substantially altering Mesoamerican concepts of the cosmos and the proper ways of dealing with its agents. Human action continued to be important for the maintenance of cosmic spacetime, and Indigenous actors found ways to adapt the Christian festival calendar to structure their "interaction with sacred beings" (Burkhart 1989, 74). The medieval depiction of saints accompanied by animals and other elements was also part of a European pictorial genre that allowed for the Indigenous depiction of beings with multiple human and nonhuman natures, although this practice was sometimes cause for suspicion.

Part of what made nepantla possible was the Counter-Reformation baroque aesthetic and ideology, which constituted one of Spain's major imports to the Americas. We will return to the critical possibilities of the baroque in later chapters. Here, it is enough to note that the "ample, dynamic, porous, and permeable" forms of the historical baroque were, ironically, themselves "colonized," eventually incorporating "the cultural perspectives and iconographies of the Indigenous and African laborers and artisans who built and decorated Catholic structures" (Zamora and Kaup 2010, 3). The eighteenth-century chapel of Santa María Tonantzintla—dedicated to the Virgin Mary of New Spain, herself an expression of nepantlismo, born from an amalgamation of the Spanish Virgin of Guadalupe and the Nahua earth deity Tonantzin, "our mother"—is a well-known example of the baroque as counter-conquest (Lezama Lima 1993, 91). Here,

> the Creole Baroque, reflected in the gilded wood altarpieces with their Solomonic columns and *estofado* saints, is undermined by the Indigenous presence . . . reflected in the multitude of Indian angels wearing feathers, armed with bows and arrows, painted with intense and loud colors, and filling the barrel vaults that are hardly visible beneath the decorative profusion. . . . Hybridity, cultural mixing, and symbiosis make the American Baroque a bizarre, fanciful, colorful, and popular art that, far from reflecting [an] experience of submission . . . in fact constitutes a vigorous sign of New World originality. (Celorio 2010, 495–96)

Rodrigo Díaz Cruz (2023, 99) writes, "Being *nepantla* implies managing diverse repertories of ritual practice and suppositions about the world that can be contradictory and in conflict, although not necessarily incompatible." At what Burkhart (1989, 187–88) calls the "dialogical frontier" between precolonial Mesoamerican and colonial European structures, Indigenous citizens of New Spain maintained many elements of their cosmological, if not political, systems, even if they were disguised as Christian.

> The upper world became peopled with saints rather than *teteo*, with the saints filling the same cult roles; the underworld with *tlatlacatecolo* [owlmen or demons] rather than the similarly peripheral underworld deities. Penitential rites were connected with the Christian calendar but continued

to serve the purpose of achieving merit. Christian purification rites functioned to remove *tlazolli* [impurity, translated to "sin"] and restore order. Salvation failed to displace the basically custodial focus of Nahua ritual and this-worldly concerns of its participants. (187–88)

Some Final Thoughts

In this chapter, starting from the postrevolutionary construction of the raza cósmica and a mestizo national identity, I've tried to pick apart the phrase "We have always looked to the stars" to show the complications of the "we" and the "looked" and the "stars" in this phrase. The implied reference to precolonial astronomical practices as a kind of protoscience that surpassed classical European astronomy or anticipated modern scientific astronomy, which is also expressed in the use of Indigenous terms for the names of technological artifacts, as we shall see, tends to obscure the complexity of the highly elaborated relational constructions, values, and practices that made up Mesoamerican cosmos.

While acknowledging the problems raised by an irreflexive appropriation of precolonial Indigenous culture, I would like to point out that the "we" in "we have always looked to the stars" is also constructed in contrast to the "they" of Euro-America, and that Mesoamerican astronomy has found a place in the archives and repertories (Taylor 2003) that nourish Mexican space milieux, alongside other actors and objects whom we will meet in the next chapters. The discourse of an ancestral knowledge that belongs to "us" and not to "them," the idea that "we were badasses," contests the neocolonialist pretensions of the U.S. flag on the Moon, as well as the universalist pretensions of the slogan "Space is for everyone."

Rather than understanding the negotiations between Indigenous and non-Indigenous knowledges after the conquest as a form of "syncretism," I prefer to see them as a process of baroque nepantlismo, a folding and unfolding of disparate aesthetic, epistemological, and practical elements from "the in-between," within the context of colonial violence and inequality. The processes of globalization that began with the colonialist projects of the fifteenth and sixteenth centuries were certainly cosmopolitan, as they attempted to incorporate alterity into "a shared universe" whose terms were set by hegemonic European discourse, but they were not always cosmopolitical, a term that characterizes the processes of "disagreements and negotiations between worlds" (Neurath 2021, 145).

The experience of colonialization put Mesoamerican notions of cosmic diplomacy to the test, as the Indigenous inhabitants of New Spain were forced to negotiate with new actors to survive. Christian Europeans imposed a "monohistory" on Mesoamerica. However, other cosmohistories have been, are being, and will be written from a place of nepantla. I will argue in the following chapters that this nepantlismo also characterizes many of the engagements between Mexico and outer space today, mediated by ambivalent geopolitical relations and temporal discourses.

NEPANTLA SPACE PROGRAM

Los Angeles, 2012. A dark sky with illuminated stars and constellations forms the background for the main image: a spaceship in the form of a ripe ear of corn. Cornsilk sprouting from the top of the ear forms lines that suggest a jet engine, and in each kernel is the masked face of one member of the Zapatista Army of National Liberation (EZLN). The Zapatista flag is nestled among the husks, which are inscribed with the EZLN's demands: land, liberty, work, health, independence, culture, education, justice, housing, food, peace, information, and democracy. Three small snails are perched on the front of the ship, each with the face of a Zapatista, a visual reference to the *caracoles*, or organizational regions of the Zapatista autonomous communities. A satellite with a Zapatista face is flanked by a crescent moon wearing a bandana and a shining sun, whose surface itself forms a mask. The satellite and the moon look down at the corn-ship, while the sun's eyes stare directly at the viewer.

The image formed part of an installation of the collective artwork *Autonomous InterGalactic Space Program* at the Roy and Edna Disney CalArts Theater (REDCAT). It was the outcome of a collaboration between the Portuguese artist Rigo 23 and "weavers, seamstresses, painters, carpenters, and cultural activists" from Chiapas. The project combines Mayan cosmology and Zapatista iconography and was inspired by the theme "Another World, Another Path," the theme proposed by the EZLN

FIGURE 6 *Autonomous InterGalactic Space Program,* an ongoing collective collaboration led by artist Rigo 23. Courtesy of Rigo 23.

at the First Festival of Dignified Rage in 2009. Its goal was "to further engender the interplanetary transmission and dissemination of the political and cultural ideology of the Zapatista movement into cosmic space and beyond."[1]

The installation was immersive, taking the form of an artisanal planetarium composed of materials like textiles, basketry, wood carvings, and painted murals, as well as videos and photographs: a Zapatista-inhabited planet and cosmos that combines images of Mayan-inflected space technology with images demanding "a world of many worlds" and promising that "another world is possible," along with slogans denouncing capitalism and neoliberalism. In the center of it all was a giant, three-dimensional version of the image described above: a wooden corn spacecraft in full color. Artists had painted upside-down baskets with the masked faces of the intergalactic community, and the pilots were Zapatista snails knitted out of wool. It felt both radically playful and deadly serious.

CHAPTER 2

MEXICO IN ORBIT

"Mexico, I Will Come on Board Your Ship!"

I met Mario Arreola, at the time the director of Science and Technology Outreach at the Mexican Space Agency (AEM), in 2018. We were sitting in his office, surrounded by outer-space objects: homemade bottle rockets; children's paintings of galaxies and spaceships; model satellites; reproductions of images taken by the Hubble telescope; and a retouched photograph of Neil Armstrong taking his first steps on the Moon, only this time planting a Mexican flag onto the lunar surface. He showed me the first-place submission in the space-art contest sponsored by the agency, exclaiming, "Here is the cover for your book!" Titled "Mexico, I Will Come on Board Your Ship!" the work features a black canvas covered in white dots: a starry night in outer space.

The eleven-year-old artist has placed the Moon in the upper-left-hand corner: a grayish-white sphere with washes of pale colors. The larger Earth, with pastel blue oceans that surround the American continent, painted in warmer greens, oranges, and yellows, occupies the lower half of the canvas. But this is all background; the protagonists of this space scene are the girl, the ship, and the satellites. The girl wears a gray-and-pink space suit with a shoulder patch—a small Mexican flag—and she clings with one hand to a ladder that undulates from its base on Earth up to the ship: a futuristic caravel like the ones Christopher Columbus's crew used on their fifteenth-century transatlantic journey. Its sails, marked with another Mexican flag and the initials of the Mexican Space Agency, billow, but it seems to be propelled by rockets with jets that shoot bright flames. The scene is also inhabited by three objects that orbit Earth: a representation of the defunct space station Skylab, a resupply ship bound for the International Space Station (ISS), and a satellite that might be an image of Solidarity, the second satellite system launched by Mexico in the 1990s. For Mario, "Mexico, I Will Come on Board Your Ship," with its colonial callbacks, national iconography, and futuristic projections would be the ideal image to grace a book that I might eventually write about Mexican engagements with outer space.

This chapter serves as an introduction to the milieu of Mexico's "space sector," that is, the network composed of humans and technology involved in the country's outer-space activities. Here, I focus on the development of space infrastructure, particularly satellites.[1] The purpose of this historical overview, interwoven with ethnographical narratives, is threefold. First, satellites play a large part in the ongoing negotiations between imaginaries of modernization and national identity, which, even in the "space age," continue to reference the idea that "we have always looked to the stars" explored in the previous chapter. Second, they provide a window into geopolitics and national governance, making visible the tensions between Mexico's technological dependence on the United States and the nation's aspirations to technological sovereignty, as well as the complex relationship between state power, private industry, and citizen initiatives. Finally, satellites simultaneously serve as practical and poetic objects. They are the fundamental elements of an "orbital technosphere" (Gärdebo, Marzecova, and Knowles 2017) that extends national and planetary borders upwards and outwards even as they depend on terrestrial infrastructure. At the same time, they function as cosmic prostheses for human bodies, affects, and subjectivities. Challenging the idea that the satellite gaze exemplifies a "view from nowhere," tracing the historical and epistemological multiplicity of satellites in Mexico allows for "holding *the planet* and *a place on the planet* in the same analytic plane," in the words of Gabrielle Hecht (2018, 112). To use Hecht's terminology, satellites become "interscalar vehicles" that can be "means of connecting stories and scales usually kept apart" (115).

FIGURE 7 *México, ¡me subo a tu nave! / Mexico, I Will Come on Board Your Ship!*
Courtesy of Mario Arreola, AEM.

It All Started with Sputnik

The *Fellow Traveler* hangs from the museum's ceiling: a round metallic ball with four spidery appendages. The National Polytechnic Institute (the "Poli") in the north of Mexico City houses this Sputnik within the Luis Enrique Erro Planetarium. According to one of the museum guides, the object was donated to Mexico by the Soviet Union at the end of the 1980s, and it is "one of the satellite's three original prototypes. It did not make it to space, but it came close." Because the original antennas were lost at some point, the extensions on display are replacements.

Annick Bureaud (2021, 79) characterizes Sputnik, the first artificial satellite, as "an all-round object (political, ideological, demiurgic, technical, performative, utopian, and aesthetic)." Its 1957 launch marked the first time that humanity (or a part of humanity) would escape Earth's gravity using technology, and it stunned the world. Viewed from Mexico, though, human activities in space provoke profound ambivalence. On the one hand, Sputnik, the effervescence sparked by the flight of Yuri Gagarin, and the Apollo Moon landings inspired a flurry of space-related activities, including the construction of ground tracking stations, launches of suborbital rockets, the consolidation of university science programs, and participation in the debates around international space-use treaties (Johnson 2020). During the 1950s and 1960s, Mexican architects designed sleek skyscrapers and imaged modernist "satellite" cities, expanding urban life outward and upward, which were advertised with cartoon space aliens to illustrate the projects' futurist aspirations,[2] while Mexican muralists depicted space exploration in hopeful terms as the logical outcome of social development tied to technological processes. On the other hand, Mexico's complicated relationship with both the United States and the Soviet Union, as well as its precarious position with respect to "modernity," created a context in which the discourse of "for all mankind" was viewed with suspicion and, in some cases, hostility.

One of the most important moments in the history of space exploration, but also for global scientific collaboration, was the International Geophysical Year (IGY) of 1957–1958. Sixty-seven countries, including Mexico, and nearly eighty thousand scientists participated in this international project to study Earth as part of a much larger cosmic system. National committees formed the basis for the planning and organization of the activities of the IGY. In Mexico, scientists under the leadership of the director of the recently created Institute of Geophysics at the UNAM participated in the establishment of new oceanographic and meteorological research stations, gravimetrical expeditions to Central and South America, the observation of solar activity (because of a peak in the eleven-year solar cycle, the aurora borealis was visible even in Mexico during the IGY), and atmospheric and seismological studies, among other projects (Urrutia Fucugauchi 1999, 128). Mexico was also designated the seat of the Pan-American Committee for the IGY by the Organization of American States (OAS).

Along with scientific advances in geo- and astrophysics, like the discovery of the Van Allen radiation belt, one of the most important achievements during this year was the launching of Sputnik 1 and 2 by the Soviet Union, and Explorer 1 by the United States, marking the beginning of the Space Race. However, at the inauguration of the IGY, U.S. President Dwight D. Eisenhower declared that "the most important result of the International Geophysical Year is the demonstration of the ability of peoples of all nations to work together harmoniously for the common good" (Kohut 2008, 30). Significantly, NASA included an Office of International Cooperation from its inception in 1958.

In Mexico, journalists reported on Sputnik's launch on the same day of the event, at the beginning of the General Motors news program; for most Mexican viewers, this was the beginning of the Space Race. A few days later, the program's lead story featured a report about Sputnik's technical elements, accompanied by images of a U.S. satellite, as the Soviets had not yet released their own images, as well as a recording of Sputnik's transmission, made by engineers at the UNAM's Institute of Geophysics. Mexican journalists began to speculate about the possible military uses of satellite technology in the context of the Cold War "balance of terror" (González de Bustamante 2012, 121).

Two months after the launch of Sputnik, in December 1957, a team of scientists from the Autonomous University of San Luis Potosí (UASLP) successfully launched the first Mexican rocket into the atmosphere from a desert region, baptized by journalists as "Cabo Tuna" or Cape Cactus Fruit, in a tongue-in-cheek reference to Florida's Cape Canaveral. Although at the time Mexico was far behind the United States in terms of investment in science and technology, two physicists who had obtained advanced degrees at Purdue University had begun to work on applied space technology at the UASLP. They started experimenting with cloud seeding, but realized they would need to develop new propulsion systems, so they modified the "classic fireworks used in religious celebrations to reach new heights" (Martínez and Palomares 2004, 6). Eventually, the Cabo Tuna engineers constructed the solid fuel rockets Zeus and Filoctetes, and "two- and three-stage rockets soaring through the sky of Potosí were common sights" (7).

Rocket launches were also reported on national television, showing "the country's concern, almost obsession, with technology as a symbol of modernity and progress . . . if the country was technologically advanced enough to launch a rocket, then it was certainly on its way to creating a more developed and modern society" (González de Bustamante 2012, 123). In 1962, President Alfredo López Mateos created the National Commission for Outer Space (CONEE), the first governmental space organization in Mexico, whose objective was to provide support for research on rockets, telecommunications, and atmospheric studies. The commission's advisory committee included representatives from the UNAM, the IPN, the SCT, and the Secretary of Foreign Relations (SRE).

Two years after the first launch from San Luis, the SCT's rocketry team also began to build and launch rockets, which they baptized with Nahuatl names. Tototl ("bird") reached an altitude of twenty-two kilometers, using solid fuel like the San Luis rockets. After its integration into the CONEE, the team then decided to experiment with liquid fuel, following the example of the German rockets built during World War II. Mitl-1 ("arrow") attained an altitude of 50 kilometers, and Mitl-2 breached the Kármán line, reaching 120 kilometers. The commission also experimented with smaller rockets, like Tlaloc, designed for cloud seeding. Other programs sponsored by the CONEE included the creation of a program of atmospheric balloon probes, the acquisition of meteorological and remote sensing data from U.S. satellites, research into space medicine, and participation in international debates around space law (Gall and Álvarez 1987, 111–14). As González de Bustamante writes, Mexico's "entry into the Space Race" was fundamental for the country in three ways. First, it strengthened Mexico's position as a leader in Latin America. "While the United States and the Soviet Union raced to put a man on the moon, Mexico, Brazil, and Argentina competed on a secondary level" (González de Bustamante 2012, 121). Second, the development of space technology, first rocketry and then satellites, was a message to the nation about Mexico's aspirations to modernity. Third, Mexico's participation in U.S. space projects fortified the diplomatic relations between the two countries.

The main NASA project that included Mexican participation was the construction of a ground tracking station in Guaymas, Sonora, which was used from 1961 to 1963 in support of Project Mercury, and later Project Gemini. Tracking stations were also constructed in Spain, Nigeria, and Australia. Mexican television broadcast the latest news on the project to show the government's efforts to modernize the country. On June 26, 1961, the day of the ground station's inauguration, the nightly news reported that Mexico was cooperating with the United States "in a project with no warlike intentions," a statement meant to appease those opponents of the project who felt that it went against Mexico's vow to "remain neutral in the face of international conflict" and others who feared that the Mercury project could be used for military purposes (González de Bustamante 2012, 125). President Kennedy enjoyed great support in Mexico, but many Mexicans decried Kennedy's policies in Cuba, especially given the Bay of Pigs invasion that had occurred only two months earlier. It is also telling that the report on the Guaymas station aired on the same day as a report on tensions between Mexican migrant workers and U.S. unions (127–28). The station was dismantled and returned to the United States when Project Gemini ended. According to Román Álvarez (1987, 120),

> there is no evidence that the station's operation (located barely 500 kilometers from the border with the United States) resulted in any technological or scientific benefits for the country. In a best-case scenario, one might say that it was an act of friendly collaboration on the part of Mexico, but in no way did it promote the development of Mexican space technology.

Despite important advances, the CONEE was plagued by internal tensions between "technicians" from the IPN and "intellectuals" from the UNAM and between the various government agencies whose interests did not always align (Borrego and Mody 1989, 267). In a co-authored article, Ruth Gall, one of the two UNAM representatives to the advisory committee, and Román Álvarez criticized the CONEE for relying too much on support from NASA and, "as is habitual, ignoring or not taking advantage of the important scientific resources that existed and continue to exist" in Mexico (Gall and Álvarez 1987, 115). This complicated context, as well as the economic crisis of 1976 and the creation of other state agencies that duplicated its functions, contributed to López Portillo's decision to dissolve the CONEE in 1977. More than thirty years would pass until the creation of another national space organization, the Mexican Space Agency (AEM), signed into law in 2010.

Mexico in Orbit

The first televised event broadcast by satellite was the 1963 launch of astronaut Gordon Cooper as part of Project Mercury (González de Bustamante 2012, 19). Four years later, during the Summer of Love, the BBC transmitted *Our World*, the first live multinational television production, using Intelsat and NASA satellites. Fourteen countries contributed to the program's content, including Mexico, one of two developing nations to collaborate on the so-called global event that was seen in twenty-four countries by around five hundred million viewers. The program, whose producers decided on the theme of population explosion as a topic that would ostensibly speak to a universal preoccupation, began with images of babies being born in four countries around the world: Japan, Mexico, Canada, and Denmark. "Edmundo," born in Mexico City, was shown "fresh, red and still unseparated from the umbilical" (Parks 2005, 29). The transmission began with a panoptic satellite image of Earth: "This is our world as no one on the world can see it" (30). Viewers saw images of a steel mill, a traffic jam, a weather station, and the building in North Carolina in which Lyndon Johnson and Alexei Kosygin were meeting to discuss world peace. Other segments addressed world hunger, the achievements of the industrialized world, and cultural and artistic diversity. By celebrating the benefits of industrial modernization and pointing to "population explosion" as the source of world hunger, a common view at the time, *Our World* presented industrialization as the solution to the unequal distribution of resources, rather than its cause. The program's concluding segment focused on the technical advancements that would allow humans to explore the universe, beyond the "limits of our world," bringing *Our World*'s "claims to global presence full circle, for the power to see and experience the earth as a unified totality brings with it the power to know and contextualize the relations of those dwelling in it" (32).

In addition to the birth of Edmundo, Mexico contributed footage to the artistic and cultural section of the program, which included scenes from rehearsals of Franco Zeffirelli's film *Romeo and Juliet* in Italy and the opera *Lohengrin* in Germany, modern art and performance in France, pianists at New York City's Lincoln Center, and the Beatles recording "All You Need is Love" in London. The Mexican contribution cut together film of folk dancers, a singing charro on horseback, images of urban modernity such as highways and skyscrapers, and video of technicians watching the replay of these same images, all set to the song *"México lindo y querido."*

Although the program was meant to be completely live to showcase the power of global satellite communications, the Mexican cultural footage was, in fact, the only element of "canned liveness" (Parks 2005, 38). The use of prerecorded images was an outcome of a desire on the part of producers to allay fears about Mexico's technological capacity, a strategy that "differentiated a core of established European broadcasters from more peripheral and tenuous contributors, especially Mexico, which was subtly displaced from the 'global now' by virtue of its prerecorded segment" (39). *Our World*, as both message and medium, located Mexico in an ambiguous space between developed and developing. Mexico's cultural segment, itself an amalgam of symbols of the past, the present, and the future, concluded with a reminder of the nation's sociotechnical desire for inclusion in global modernity, as the announcer invited the international public to witness the Olympic Games that would be televised to the world by satellite from Mexico City the following year, the first Games to be held in a Latin American country.

On October 2, 1968, hundreds of unarmed students, many from the UNAM and the IPN, were massacred by military forces in the Plaza of the Three Cultures in the Mexico City neighborhood of Tlatelolco, while thousands more were injured. The students, supported by many of Mexico's intellectuals, had been marching peacefully, protesting the use of public funds to finance the 1968 Summer Olympics to be held in Mexico, shouting "We don't want Olympic Games. We want a revolution!" (Poniatowska 1975, 12). Protesting the Olympic Games was one of the most visible elements of a broader student movement centered on resisting government authoritarianism. Demands included effective democracy, university autonomy, labor reform, and an end to corruption and police brutality. In the aftermath of the massacre, the government jailed thousands of protesters and "suspected dissidents," and many were tortured. Alongside the Mexican army, the main perpetrators of this horrific act of state violence were the members of an armed group called the Olympia Battalion, which was ostensibly created to guard the installations that would be used for the Olympic Games but in reality was a paramilitary team directly answerable to the presidency of Gustavo Díaz Ordaz. Ten days after the massacre, the Games went ahead as planned. Vicente Saldaña, an engineering student at the IPN, remembered:

> There were a number of us, of course, who claimed that we students had to take advantage of the Olympics, of the huge crowds that would attend them, to attract attention to our problems, and we realized, naturally, that we'd be the sour note, the blot on the image—like when the president visits some little town and among the banners reading WELCOME and MANY THANKS there's one that says OUR TOWN HAS NO WATER AND NO ELECTRIC LIGHTS. We were the one voice that was off-key in the universal hymns of praise, but that's a far cry from wanting to sabotage the whole celebration—not to mention actually succeeding in doing so! What's more, there were a few kids who knew what the score was, but the rest of them were just a bunch of sheep . . . there were lots of kids, lots of Movement people, who either actually attended the Games or at least watched them on television. It was revolting! To think they could watch the Games like that—over the dead bodies of their comrades and the thousands of people who'd suddenly disappeared, people we knew had been thrown in jail but had had no news of since. (146)

The Mexican government had affiliated with the global satellite network Intelsat to facilitate the transmission of the Games. The SCT also built Mexico's first satellite tracking station in Tulancingo, Hidalgo, the largest satellite antenna in the world at the time. Mexico became the second country to transmit the Olympic Games by satellite and in color.

In 1972, Mexico began to buy images produced by NASA's Landsat system, the first series of satellites designed to monitor natural resources through the production of visual data. However, renting communication satellites and buying Landsat images from NASA for Earth observation proved to be a short-term solution, given the increasing importance of satellite data from the latter half of the twentieth century. International collaboration and sharing of images had served its objectives, like the successful eradication, thanks to remote satellite observation, of a screwworm plague that had attacked Mexican livestock, but there were concerns over the consequences of allowing foreign countries access to Mexican data. In September 1971, the Mexican government demanded that the United Nations agree that "no data would be collected over Mexican territory from the air or space without prior permission" (Mack 1990, 187).

Worried about this lack of control over national data and the telecommunications gap between urban and rural populations, the Mexican government negotiated the obtention of two geostationary satellite positions in 1979. The national satellite project was also supported by the communications company Televisa ("Television Via Satellite"), as the colocation of satellites in the newly acquired positions would allow for the transmission of its television programs throughout its principal markets: Central America, the Caribbean, and part of South America. Although their concerns were largely economic, both Televisa and the government publicized their efforts as part of an attempt to resist the "Americanization" of Mexican values through the production of regional and national television content. The project was also publicly justified by appealing to the possible social uses of the new satellites, although no major social agency participated in the decision-making process (Borrego and Mody 1989, 271).

Three years later, at a cost of ninety-two million U.S. dollars, the SCT hired Hughes Space and Communications (later Boeing) to construct Mexico's first satellite system, consisting of two satellites, Morelos 1 and 2, and a ground control center in the Iztapalapa area of Mexico City. The state-run company Telecomm would take charge of the satellites' operation, but the design and delivery of the new technology, as well as training and maintenance, would be undertaken by international experts. This strategy was the subject of critique by some, as it meant that

Mexico would continue to be technologically dependent on the United States instead of moving toward national sovereignty and autonomy, unlike countries such as Brazil and India, which invested in the longer-term strategy of consolidating national human and technological resources (Borrego and Mody 1989). In 1985, through an agreement with NASA, the Mexican government put its two satellites in orbit and sent Mexican citizen Rodolfo Neri Vela into space on board the space shuttle Atlantis.

The SCT promoted the satellite system with a lavish documentary narrated by the distinguished actor Claudio Brook, who had appeared in numerous Mexican and international films and television productions. The documentary's unimaginative name, *Morelos Satellites System*, belies its stimulating content, combining fonts, sound effects, and animations in the style of an eighties video game with references to Mayan astronomy and explanations of how satellites work. The director shot Brook at Chichén Itzá in Yucatán, Cape Canaveral in Florida, the Johnson Space Center in Texas, and the Hughes factory in California. The actor explained that the satellite would "be controlled from Mexican territory," and "would be in the hands of Mexican technicians and engineers." It would benefit rural communities through remote access to education, as well as supporting the modernization of the Federal Commission of Electricity (CFE), Mexican Petroleum (PEMEX), the health sector, and, of course, telecommunications companies that would be able to contribute more to the gross national product. He traced the development of technology from tools made of stone, wood, and bone to "powerful instruments" like satellites:

> When the two Morelos reach their geostationary orbits, the celestial dome will contain two new heavenly bodies, and two earthly satellites, that the Mayans would probably have believed to be two new gods. The sky then wasn't the same as it is today. It is easy under a starry sky, when we find ourselves in someplace without light or contamination, to raise our gaze, and, between the rapid movements of comets and the slow movements of stars, to discover the agile but uniform navigation of a small, luminous spark. It is a satellite. And there are so many in the nocturnal firmament that we can find several of them in just a few hours of observation. (Adalid 1985)

A reference to the Mayan *Popul Vuh* concluded the film: "We are the avengers of death. Our line will never be extinguished so long as there is light in the morning star" (Adalid 1985).

An Astronaut and an Earthquake

The second Latin American in space, after Cuban cosmonaut Arnaldo Tamayo Méndez, was Rodolfo Neri Vela, born in the state of Guerrero, and a specialist in antennas and satellite communications, chosen in a national competition to join the crew of the space shuttle Atlantis as a payload specialist. The SCT launched the call for participants in a "contest for the selection of a passenger to fly on the space shuttle" at the beginning of 1985, as well as a "contest for the undertaking of an experiment in space." The poster announcing the two contests proclaimed, "We are bearing in mind [in a play on words, '*tenemos presente*'] our future." Among other requirements, astronaut candidates were expected to have advanced science or engineering degrees, professional experience in research or teaching, fluency in English, and excellent physical condition (Neri Vela 1993). Of the more than twelve hundred candidates who registered, ten finalists were selected, including the eventual winner Neri Vela, at the time a professor of engineering at the UNAM; his alternate Ricardo Peralta y Fabi; Francisco Javier Mendieta, second alternate, who, years later, would become the first director of the Mexican Space Agency; and only one woman, oceanographer and marine biologist Vivianne Solis Wolfowitz.[3]

By the 1980s, the ruling Revolutionary Institutional Party (PRI) had dominated Mexican politics for decades, and elections had become rituals of legitimation rather than the expression of popular desire or the outcome of vigorous debate in a public sphere (Lomnitz-Adler 1996). The system was complex and corrupt, and the opaque processes through which power flowed were (and still are) subject to a wide range of popular interpretations. The space passenger contest, organized by the SCT as part of its agreement between the Mexican government and NASA to send a series of communication satellites into orbit, was described by some of my interlocutors at the AEM, especially those belonging to Neri Vela's generation, as a political rather than a purely technoscientific affair.

Several of these contacts participated in the space passenger contest or collaborated with the astronaut as colleagues or students in the years after his flight and so had first-hand memories of the process. Rumors and conspiracy theories abound. In one conversation, the name of one of the candidates chosen as an alternate came up. He had great qualifications, they said, but he did not have a *padrino*. They told me of journalists who "discovered" secret fraternities within the PRI through which power and favors circulated and which could have influenced the contest's outcome. "That's how things were, *ni modo*," they concluded with a shrug.

However it happened, Neri Vela was selected as the winning candidate, and, as a payload specialist, he was tasked with overseeing the launch of the satellite Morelos 2 (see chapter 3), as well as undertaking research into "bacterial growth, the germination of various seeds, the transport of nutrients through plants, the validation of theories of 'electropuncture' to stimulate disequilibrium in human organs and, of course, photography of Mexico and its recovery in the aftermath of the earthquake" (Evans 2012, 427).

In September 1985, just two months before the planned launch date, a magnitude 8.1 earthquake hit Mexico with catastrophic consequences, especially in Mexico City, built on an ancient, unstable lakebed. Although the official death toll was 3,192, President Miguel de la Madrid had ordered a news blackout, obscuring accurate accounting. At least twenty thousand people perished because of the collapse of poorly constructed buildings in the country's capital. At least two hundred fifty thousand people were left homeless (according to official figures—the true number is likely to have been much higher), and the economic cost is said to have ascended to eight billion dollars. The government of de la Madrid was severely criticized for its response to the disaster, and in the absence of institutional action Mexican citizens (particularly those not aligned with the PRI) were forced to organize themselves to deal with the disaster. Indeed, the 1985 earthquake was a defining moment in the creation of an active civil society in Mexico.

Mexico's communications system was among the earthquake's casualties. The SCT's main office, located in downtown Mexico City, collapsed, bringing down with it most communication between Mexico City and the rest of the world. The installations housing Televisa also collapsed, killing eighty of its workers. The disaster left the ground station that transmitted signals from the Morelos 1 satellite that NASA had launched into orbit in August of that same year without electricity. Neri Vela (1992, 149), emphasizing the importance of satellite communication, comments that, fortunately, Imevisión, a television station transmitting from southern Mexico, was able to broadcast its signal through a hastily cobbled-together relay to one of the parabolic dishes that communicated with the satellite, and therefore retransmit Televisa's news reports nationally. Vanessa Freije (2020, 139) calls the 1985 earthquake "Mexico's first mediated natural disaster," a phrase that calls attention to the irony of televising the collapse of subpar urban infrastructure using newly minted space infrastructure.

As Mexico City chronicler Carlos Monsiváis (1986, 19) reported, not all viewers were impressed with national media coverage of the tragedy and its "back to normal" messaging.

> The unctuous announcer appears in the frame and invites us to accompany him as he walks through the eternal city. Look, the Angel of Independence fell in 1957, that is true, but look how stoic she is now. The announcer frowns intelligently, composes his face in the proper direction to reassure the housewives, and points on a map of the city to the affected portion. "Do you see? It's a tiny piece." What a shame that he hasn't gotten his hands on a map of the cosmos. What are, exaggerating, four thousand buildings and twenty thousand dead in the context of galaxies?

Mexico's participation in the NASA space mission was not a national priority in the earthquake's aftermath and was in danger of being postponed until communications could be re-established. However, the project did go ahead, and the satellite Morelos was placed in "parking orbit" until the telecommunications system could be restored. The mission STS-61-B launched on November 26, 1985, from Kennedy Space Center; the shuttle and its seven-person crew (six U.S. astronauts and Neri Vela) spent just under seven days in orbit around Earth. Neri Vela and his alternate had trained at the Johnson Space Center for five months before the launch, which may not have been long enough to convince his commander Brewster Shaw of his preparedness. Neri Vela remembered (1992, 188) that Shaw kindly told his parents, who had traveled from Guerrero to see the launch, not to worry, that all would be fine and that "Rodolfo is fully trained, and we're glad to have him with us." But Shaw later recounted "I'm probably a paranoid kind of guy, but I didn't know what he was going to do on-orbit, so I remember I got this padlock and . . . went down to the hatch on the side of the orbiter and I padlocked the hatch control so that you could not open the hatch. . . . I don't think Rodolfo noticed it but some of the other crew noticed it" (Evans 2012, 424). There was no "freaking out" on Neri Vela's part, but as it happened the padlocking of the hatch control became standard shuttle procedure.

During his time in orbit, Neri Vela carried out the three winning Mexican experiments, studying the effects of electropuncture on the human body in space (using himself as a test subject), the influence of zero gravity on the sperm of a stallion, and the germination process of amaranth, wheat, and lentils in space. Neri Vela described the fifth day of the voyage as "a Mexican day." In the morning, NASA awakened the crew with the "*Canción mixteca*" and later, shortly after finishing the fifth stage of the germination experiment, Neri Vela spoke with President Miguel de la Madrid. "In my name and the name of all Mexicans," de la Madrid greeted him, "I send you an affectionate salute. We are very proud of how you have undertaken this historic mission of being the first Mexican in space. . . . How do you feel?" Neri Vela answered "Very well, Mister President. It is a great honor for me to be able to greet you from NASA's shuttle Atlantis, and I hope to send a warm greeting to the people of Mexico, given that it is thanks to the efforts of every one of my countrymen that, at this moment, Mexico is represented around our marvelous planet. I have been able to see . . . the beauty of Mexico, from the peninsula of Baja California to Yucatán and Quintana Roo" (Neri Vela 1992, 232).

The official photo of the 61-B crew features four members in flight suits and two in space suits, posed in front of a U.S. flag. But another, more playful, photo was taken unofficially. It featured two crew members in lab coats (one with a bowler hat), two in space suits and construction hats (one with a toy kangaroo in his lap, referencing an Australian satellite that would be placed in orbit), one in a vintage aviation suit, the commander with a baseball cap and a "Boss" badge, and Neri Vela in a giant sombrero and serape. According to crew member Sherwood "Woody" Spring (the one holding the stuffed kangaroo), "That photo could not be the official one, because the Mexican government took a little bit of umbrage at Rodolfo being dressed up in a serape and a sombrero, but then post-flight we went down to Mexico City . . . and the first thing they did was take us to the folk ballet, where everybody is dressed up exactly like that!" (Evans 2012, 424). Neri Vela took the sombrero picture in stride, writing later that "at least in his case, he never felt any kind of discrimination. On the contrary, he would always have happy memories of how he had been treated in every moment, during the several months that he had represented his country in a strange land" (Neri Vela 1992, 263). The Mexican government may not have been thrilled to see its representative stereotyped in NASA photographs, but it did support his request to take tortillas, a national symbol and staple of the Mexican diet, on board the shuttle. NASA approved the request as well, and now tortillas are a mainstay of astronauts' diet, given that they do not spoil quickly (thanks to the development of shelf-stable tortillas by Taco Bell scientists), nor do they disintegrate into crumbs that could wreak havoc in a low-gravity environment (Pisano 2024).

Outside of Mexico, STS-61-B is notable for being NASA's last shuttle mission before the Challenger disaster of 1986. But for many Mexicans, the mission was meant to mark the beginning of a new era for Mexico in outer space, a chance to "reach a relevant place among the diverse countries of the world" and achieve "social and technological" progress for its population (SCT, n.d.). Neri Vela (1992, 197) described the response by the Mexican public on the day of his flight and its coverage in the national press: "It was obvious that the entire population of the country knew about what would happen that day, historic for the nation; finally, a Mexican citizen would stride into space, carrying its flag and putting it before the eyes of the world."

Techno-Nostalgia

In 1993, the SCT replaced the Morelos satellite system with the Solidaridad system, also constructed by Hughes, for three hundred million dollars. By this time, satellites were too commonplace to require the production of propaganda films such as the one described above. The government privatized the telecommunications industry in 1997, and its satellite systems came under the control of the Mexican company

SATMEX, which sent another generation of satellites into space. The transnational company Eutelsat acquired SATMEX in 2014 for 831 million dollars (in addition to the 311-million-dollar debt that had to be paid) and renamed its fleet of satellites. In 2005, the SCT concessioned one of its geostationary slots to the national company QuetzSat, which launched the Quetzal 1 satellite in 2011. The SCT, hoping to bolster national security, acquired the system MEXSAT (not to be confused with the private SATMEX), with three satellites launched between 2012 and 2015: MEXSAT 1 (also called Centenario), which was destroyed during a launch incident; MEXSAT 2 (also called Morelos 3); and MEXSAT 3 (also called Bicentenario).

In summer 2023, in an office building adjacent to the Iztapalapa tracking station, my friends Ilana and Rodrigo of the art collective TRES and I sat down with engineers and managers at Eutelsat to talk about the history of satellites in Mexico. Uninterested in the rocket launches of the 1960s and 1970s, they told us that the original Morelos system was "Mexico's first real contact with space." But satellite technology has changed during the last few decades. Morelos was like a "Rotoplás" (a brand of domestic water tanks), a giant aluminum-alloy cylinder with honeycombed panels, whose onboard computer had data buses but no central processor. "It was kind of like a single-celled organism, with no central nervous system. It reacted to its environment through irritability." Things started to change with Solidaridad, because its processes were more automated, and today's satellite technology is much more complex and sensitive. But somehow, they said, even though satellites today are more technologically impressive, "Morelos was nobler." Engineers on the ground got to get their hands dirty, because older satellites' operation required much more interaction. You could get involved in their programming; you had much more participation in their functioning. They did everything by hand, even the process of checking their positions and making sure they would not collide with other satellites. Morelos was like an old car, like a Volkswagen Beetle. It let you learn. But today everything is black-boxed, and you can't see codes and sources anymore. You can modify the satellite's footprint now, though. With Morelos, "all the satellite could see was Mexico," because the antenna's cone was shaped like the country's outline. And eventually, it could only "see" out of one "eye" after its other main sensor was damaged.

When I asked about the criticism that these satellites can't be considered "Mexican" because they were fabricated in the United States, the engineers made it clear that, in some ways, older satellites were more Mexican because there was much more local involvement in their operation, whereas now, satellites come in "kits."[4]

Satellite Dreams in Miniature

The twenty-first century saw the beginning of the era of NewSpace, marked in Mexico by two events. The first was the creation of the Mexican Space Agency in 2010, which established a representational institutional figure with the capacity to negotiate agreements with other international organizations and space agencies. In contrast to the presidential decree that brought the CONEE into being, the AEM was constituted through a legal process that involved approval by both the Senate and the House of Representatives. The second watershed event was the hosting of the International Astronautical Congress in Guadalajara in 2016. At this venue, Elon Musk unveiled his plans for the human colonization of Mars, inspiring the next generation of young Mexicans with a passion for outer space. (Not everyone was a fan, though. "That guy," one Eutelsat engineer complained, "just launches things because he can.")

The amount of economic and technological resources needed for the operation of national space programs has been a major impediment to the incorporation of countries from the Global South in space exploration and commercialization projects. The combined budget of all Latin American space agencies represents about 2 percent of NASA's budget, or 6 percent of the European Space Agency's budget. Therefore, none of these countries has the capacity to invest in expensive technological projects. But according to my interlocutors in the space sector, the transformation of the space industry projected for "emerging countries" will depend on three factors: the "democratization of space," the reduction of the costs of space technology that is resulting from processes of miniaturization, and the private sector's increasing presence in the space industry.

Despite the "democratization of space," Latin American countries continue to have little influence on or participation in the megaprojects

planned for the exploration of the solar system. As several of my interlocutors have stated, "The floor is uneven." However, they agree that low
Earth orbit has become an important area for expansion, especially given
the development of microsatellite technology, with much more minimal requirements than the massive satellite infrastructures that underlie
systems such as Morelos and Solidaridad. The Mexican Space Agency,
therefore, has set its sights on the construction of CubeSats, which are
literal "black boxes": small, standardized satellites that weigh one kilogram, contain a volume of one liter, and measure ten cubic centimeters
per unit. The development, launch, and operation of a CubeSat tend to
cost a fraction of that of a large satellite, and the price goes down when
various satellites share a rocket's payload space. The standardization of
the design also allows a developer to buy most of the satellite's components off the shelf and adapt them to the project's requirements. For this
reason, aside from their scientific value, CubeSats are also considered
to be valuable teaching tools for students and citizens (Pang and Twiggs
2011, 50). In fact, one person I interviewed at Eutelsat called CubeSats
"the Holy Grail" of university engineering programs.

The most basic CubeSats do not do much more than Sputnik did six
decades ago: they emit a "beep," which is transmitted by radio to confirm
that they continue to function in orbital conditions. But missions may
also be more complex, complying with objectives that can include detect-
ing changes in Earth's magnetic field, testing advanced technology, mea-
suring climate change, observing atmospheric layers, and imaging Earth's
surface, among other things (Pang and Twiggs 2011, 50). Although the
quality of CubeSat communication systems cannot compare with those
of larger satellites, their potential increases when they are designed as
part of a satellite constellation. Because of their size, they can be programmed to automatically deorbit when their functions cease and burn
up in Earth's atmosphere, instead of contributing to orbital space junk.
Advocates of CubeSats have emphasized their utility in the NewSpace
economy, given that both risk and investment are relatively low (50).

For my interlocutors in Mexico, however, the importance of CubeSats
goes beyond their pedagogic, scientific, economic, or ecological benefits: they seem to promise a future of independence from the hegemonic
centers of power, a kind of epistemic sovereignty (Litfin 1999) that may
assure national control not just over territorial borders and natural re-

sources, but also over knowledge and information produced in and about the nation, a latent worry since the days of Landsat. It is argued that the "development of talents"—a phrase I heard many times—that is understood as a positive outcome of the development of CubeSats is another factor that, in the long term, will help achieve this autonomy.

"Satellites are noble," a young engineer told me in 2018, five years before I would hear the same phrase from an older engineer at Eutelsat, "they are agnostic." These "invisible friends high up above us" provide a wealth of benefits, he and other participants in the Mexican space sector argue, including connecting people, rendering visible a variety of processes that occur on Earth's surface, and contributing to the dream of seeing a greater Mexican presence in outer space. Three recent CubeSat projects have competed for the title of "first Mexican nanosatellite." The first was Painani-1, developed at the Ensenada Centre for Scientific Research and Higher Education in Baja California (CICESE), at the request of the University of the Mexican Army and Air Force.

The private company Rocket Lab launched Painani-1 from New Zealand in June 2019. According to the Mexican secretary of national defense Painani-1 was merely meant to be an educational project for CICESE students, as well as a rehearsal for the development of Earth observation technology; however, given military involvement, the government asked Rocket Lab not to publicize the mission. Although the military has not been the primary user of satellite technology in Mexico, the armed forces may seek a stronger space presence in the future. In an article for the journal of the Mexican Navy, for example, Rivera Parga (2017, 33) argued for the military importance of satellites to "increase the national power of the Mexican state." After enumerating the social, economic, and political benefits of satellites, the author appealed to the rights of each state to "defend its sovereignty, not just within the terrestrial atmosphere, but outside it," given that, "if we as a sovereign State do not exercise sovereignty in our own air space and outer space, who will?" (57). In the last decade, the Mexican government has turned to satellites and other surveillance technology (like facial recognition technology, drones, and robot dogs) as part of a strategy of "hypervigilance" and data collection about groups that they consider "obstacles to national security": criminal organizations, but also immigrants and members of social movements (see Méndez-Fierros 2023).

The second Mexican satellite launched in the twenty-first century was accompanied by much greater national press coverage than the secretive Painani project. Students and professors built AzTechSat-1 at the Popular Autonomous University of the State of Puebla (UPAEP), with technical support from NASA experts. The satellite, whose name combines a recognition of the pre-Hispanic past with the appropriation of modern technology, was a one-unit CubeSat designed to test communications between small satellites and commercial satellite constellations belonging to the company Globalstar. AzTechSat-1 was launched to the ISS from Cape Canaveral on a SpaceX Dragon capsule in December 2021, contributing one kilogram to a payload of 2,585 kilograms that included genetic experiments from NASA, Anheuser-Busch experiments on the germination of seeds in space, and an ISS experiment on the use of spectrometers to detect gas leaks in space. AzTechSat-1's short life ended when it unexpectedly deorbited, disintegrating upon contact with Earth's atmosphere.

The AEM's YouTube channel broadcast the launch, retransmitting NASA's live feed. One public official who spoke at the event characterized AzTechSat-1 as "the first nanosatellite made in Mexico and probably, depending on how it is defined, also the first satellite made in Mexico."[5] The debates around satellite technology's "true nationality" continued; several of my interlocutors mentioned that AzTechSat-1 was bought as a "kit" from the United States, although its payload design and assembly took place in Mexico. But, for José Valdés, the coordinator of the Space Studies Program at the UNAM, the experience "served as training" and was an important step for the "development of talents" and the achievement of "technological independence" that would eventually "let us see what we want to see." He talked to me about the need to be able to observe "our own territory, our forests, our urban growth," track earthquakes, diseases, and volcanic eruptions, and improve communication. In the future, he concluded, these "transformative technologies" and the information they provide will generate "well-being and development." In collaboration with NASA, the AEM began to develop AzTechSat-2, a constellation of CubeSats designed to track the movements of marine mammals off Mexico's Pacific coast using satellite telemetry. However, promoters of the space industry may be unduly optimistic about satellite technology. As Australian archaeologist Alice Gorman argues, countries with low GDP "are in this position because of systemic and historic inequities relating to colonialism and capitalism. . . . Additional Earth observation data isn't going to fix global inequalities, without radical political change, too" (in Howell 2020).

Satellite Poetics

NanoConnect-2 was developed in the UNAM's Space Instrumentation Laboratory (LINX) with a payload designed to test technology in low Earth orbit and connectivity with Earth monitoring stations. In February 2021, the Indian Space Research Agency (ISRO) placed NanoConnect-2 in orbit at a height of 504 kilometers. As was the case for AzTechSat-1, this satellite shared its flight on the Polar Satellite Launch Vehicle with international projects, on this occasion as part of a package whose main payload was the Brazilian satellite Amazonia-1. According to Gustavo Medina Tanco, the head of LINX, NanoConnect-2 (unlike AzTechSat-1) is "100 percent Mexican in its technology, design, and conception; it was made on our own initiative, with our students." Of course, "100 percent Mexican" is something of a misnomer, given the international provenance of many of the satellite's elements and the number of intricate international agreements required to launch it.

I visited the laboratory in August 2019, as various members of the NanoConnect-2 team were checking the satellite so that it could undergo the battery of tests needed for launch. In my notes, I wrote "objects, objects everywhere, a bit of chaos": the lab was filled with containers, computers, printers, boxes, cables, and a large quantity of equipment I couldn't identify. Students in white lab coats worked in front of screens, doing calculations, using specialized software to analyze information, or adjusting instruments. Near the entrance was a small darkroom, with a device that had something to do with photons, and a series of 3-D printers. At the front of the lab, Medina Tanco explained with pride, was a piece of equipment that cost several million dollars, used to test instruments in extreme, space-like conditions.

During the time I observed the lab's activity, the team encountered some problems: first, the satellite wouldn't turn on, so it was connected directly to a power source. Then NanoConnect-2 had to show it could transmit and receive information. After various attempts and the use of several different computers, communication was established. The next step was to acquire liquid nitrogen for the cooling tests. I felt the intense pressure on everyone's time: "This has to happen today." Conversations focused on the resolution of concrete problems; no one was debating the philosophy, the ethics of space exploration, or the social uses of technology (at least while I was there). But there was no lack of emotion. A few days after this visit, I received a short video through WhatsApp, in which various lab-coated students surrounded NanoConnect-2, seen without its shell, its innards exposed. A green light blinked. I heard "5-4-3-2-1 . . ." and jubilant shouts when the satellite's antennas deployed with a whip-like movement.

After its successful launch, NanoConnect-2 communicated effectively with the ground station located in the laboratory. And, as of the time of this writing, the satellite lives on, as Eduardo, one of the students at the LINX, told me. He has been a satellite fanatic for a long time; he has mounted do-it-yourself antennas on the house where he lives with his parents so that he can "hunt" satellites. In this way, he says, "You can be in space, even if it's just through satellites." Eduardo built his antennas with "things you can buy in your neighborhood hardware store," and, following YouTube tutorials, he attached them to his house and combined them with a radio and a cheap computer to create his own ground monitoring station, a necessary part of a satellite's infrastructural assemblage. One afternoon, chatting in an outside courtyard at the UNAM, he showed me how to follow a satellite using an online platform that allows anyone to access the observations undertaken by members of the network of ground stations that belong to the SatNOGS project.

NanoConnect-2 appears on the platform as an image that represents the transmission of its signal, a long rectangle in tones of blue and green: "the waterfall." The wavy neon green lines represent the tracking of the satellite during the period of observation. Other, more tenuous lines cut through the image from top to bottom: "noise" in the transmission. Data can be displayed as numbers or heard as audio: a high-pitched squeal with slight changes in intensity that makes me think of bats and echolocation. Sounds appear on-screen in purple waves, as if they were tracing the satellite's frenetic heartbeat, proof that "it's still alive."

However practical a satellite might be, and however much it promises to fulfill dreams of progress and technological inclusion on diverse scales, its significance is not limited to its instrumentality. The poetic potential of satellites started to come into focus when, in 2019, Medina Tanco sent me the image of a multicolored parabola that I couldn't at first identify: a graph tracing NanoConnect-2's potential orbit.

> I see my satellite revolving in space—I'm calculating orbits to see the level of insolation the solar panels will receive in relation to time—and I imagine it up there, where I always longed to go . . . turning in the darkness of space, looking at the stars, and the continents, in an endless dance, immersed in the most absolute silence . . . tiny in the immensity of the universe.

Those involved in the Mexican space industry repeatedly state what they consider to be the benefits of satellites: fomenting talents, testing space technology, driving economic and social development, connecting people, and observing Earth processes. But, at least for those intimately involved in their creation, satellites exceed their practical functions, becoming human prostheses in space, irreducible to technophilic fetishes (although they are also that). For some, satellites are a means of achieving cosmic dreams. Medina Tanco's satellite is the consequence of "my desire to navigate infinity . . . to fuse myself with the darkness of the cosmos . . . of my crazy mix of angst and fascination when faced with the emptiness that defines our ephemeral and absurd existence."

"Space is a blank canvas," one young engineer told me in 2018, enthusiastic about what he considered the infinite possibilities of space exploration. But, for artists that have worked with satellites, "space is a *black* canvas," not just a node in a chain of data production or communication, but rather a reflection of the desire "to create something on the scale of the cosmos by placing a new celestial body within it" (Bureaud 2021, 80). And as the product of citizen science, artistic satellites highlight the process of their own creation outside institutional channels. They defy the notion that access to space should be restricted to commercial or governmental actors; therefore, the construction of an artistic satellite is both a political and a poetic act.

According to Juan José Díaz Infante, artistic satellites "change the conversation." He explained, "What I do is generate frames of thought around space and how to see the world from above." The satellite Ulises I began its life in Diaz Infante's imagination in 2010, partly in the context of the bicentenary of Mexico's independence from Spain, and partly because the artist was turning fifty and, in his own words, "needed a re-engineering." Not long after, he found an article in *Scientific American* about citizen satellites (Pang and Twiggs 2011), and so he decided to create a citizen space agency called the Mexican Space Collective (CEM) as a challenge to governmental and industrial hegemony and a means of including other visions and voices in space, an action that he felt was particularly important to undertake as a Mexican citizen. The collective built Ulises I to appropriate "the poetic energy of Sputnik. The beep as a poem, a pixel as a masterpiece. This small object the size of a basketball, this little sound, became a trigger of the imagination" (Garciandía 2017, 31).

At first, the artists who collaborated on the project thought of it as an homage to soccer (a universal language), meant to evoke teamwork and a triumphal spirit, but the sports metaphor was gradually pushed aside. The collective work eventually included pieces ranging from visual art to poems to musical compositions, and the transmission of the satellite's signal would translate to "I love the road" in Morse code. The team exhibited the project in art festivals around the world, but the practical engineering, as well as the process of financing the satellite's launch, was more complicated. After a series of intricate negotiations with Mexican and international institutions, Díaz Infante attempted to launch Ulises I into the stratosphere during the Guadalajara Book Fair in 2015 in collaboration with the LINX, but the signal was quickly lost. The artist is currently developing other versions of Ulises.

The Afterlives of Orbital Technology

The engineers at Eutelsat unwittingly echoed Heidegger's realization that tools tend to be invisible when they work properly (when they are "ready-to-hand") but become objects of attention when they break (becoming "present-to-hand") (Harman 2010, 18–19). Satellites, they told us, only become "visible" when they "misbehave." Satellite technology is even more invisible than most infrastructure, as the objects themselves orbit far above the spaces of human daily interaction and rarely enter the consciousness of their users unless the second-order technological objects that depend on them fail. The visibility of satellites is naturally more of an issue to their operators, who are intensely attuned to the moments of visibility caused by their failure, or potential failure. This visibility may come about because of the effects of solar wind on the satellites' rotation, a technical failure, or because of an "encounter" with another satellite. At Eutelsat, they respond to at least one or two "encounters" between their satellites and neighboring satellites each month. Elon Musk, they said, probably reports seventy or eighty thousand in the same period, "so it's all relative."

"Encounter" is a benign term for the "ellipsoids of error" that signal potentially violent collisions that create much of the debris found in Earth orbits, as objects in space travel at high speeds: three kilometers per second in geostationary Earth orbit (GEO) and seven kilometers per second in low Earth orbit (LEO). The Mexican Air Force tracks bits of space junk larger than eight centimeters and sends daily collision warnings. It's a kind of space junk meteorological report, they say. And there are more and more alerts, which means there is more and more debris. Space junk has become normalized; since it can't truly be cleaned up, you must learn to live with it. The only thing you can really do is clean up the orbit of your satellite by ejecting what no longer serves higher up. The GEO must be protected, they acknowledged, because it's an important economic resource, and it's nonrenewable. But cleaning up an orbit isn't enough. And it's irresponsible, complained one engineer, to keep launching things into space just because you can. If you make it, you have a responsibility to deal with its end-of-life. Like small satellites that don't do anything but repeat what Sputnik did sixty years ago. "Some countries" ("without naming names," they said) are also at fault, as they do what they like in space without informing anyone. Space needs to be regulated better, they concluded. There are holes in the legislation. Space is like a neighborhood, and what happens to one satellite affects all its neighbors. "It's like saying that the garbage they throw away in Monterrey doesn't affect us in Mexico City. But it does." And, of course, some neighbors are more problematic than others. The AstriaGraph program at the University of Texas uses filters to show which countries have more objects (including space trash) in orbit. Filtering for the United States or Russia, for example, shows a bright cloud around Earth. Filtering for Mexico shows a few hard-to-see white dots.[6]

That said, like the satellites from which they originated, pieces of space junk are largely unseen until they fall on someone's head or, as happened in 2000, in someone's cornfield. That year, in a small town in Baja California, a large metal object that rather resembles Sputnik fell out of the sky. It is now on display in the local museum, along with other objects of historical significance.[7] More recently, SpaceX has come under fire in Mexico because of space junk caused by the explosion of its Starship 36 rocket launched in May 2025 from Starbase in south Texas. The debris was found to have caused damage to vegetation and the marine ecosystem in Mexican territory (Miranda 2025).

A critical view of the material expansion of the terrestrial atmosphere seems to be at odds with other critical and ecological uses of satellites, such as Juan José Díaz Infante's art satellite Ulises, the AEM's plans to launch the AzTechSat-2 satellite constellation to track marine fauna, or local communities' use of GPS to defend their territories from the threats posed by mining, logging, and other extractivist projects. From another perspective, Annick Bureaud questions the idea that the skies above Earth should be accessible to whoever wants to occupy them. "Do we have the right to put objects in other people's skies, even for pacific and cultural purposes, without asking them first? To whom does the sky belong?" (Bureaud 2021, 80).

One can still trace the Morelos and Solidaridad satellites on several space-object tracking sites, such as www.n2yo.com, which I used to visualize the orbital figures made by the defunct objects. Morelos 2 is currently orbiting far above the Indian Ocean, while Morelos 1 is somewhere over Indonesia. Their Payload Assist Modules, upper-stage motors used to propel them to GEO that then detach from the satellite body like "cutting the kernels off an ear of corn," are also in orbit, about thirty-four thousand kilometers above Earth; one follows an arc from South America to Africa to Australia, while the other makes a similar arc at a lower altitude. Solidaridad 1's orbit forms a figure eight over the Pacific, just west of Mexico, while Solidaridad 2 traces a giant *N* over equatorial Africa. Now all four dead satellites form a part of the giant sea of space junk that encircles Earth as, in fact, Claudio Brook had explained in the 1985 documentary about Morelos. The aging ground infrastructure that supported the first satellite transmissions in Mexico is also in danger of becoming terrestrial space junk. However, in the case of the Tulancingo ground station, scientists have produced an ingenious solution. As I write this chapter, in February 2025, the antenna Tulancingo 1 is in the process of being transformed into a radio telescope, meant to eventually participate in an international network of radio telescopes that will provide information about the furthest reaches of the universe, and to serve as a model for the sustainable conversion of outdated telecommunications infrastructure to modern astronomical instruments in other Latin American countries (Manilla 2024).[8] Instead of linking Mexico to satellites to bounce terrestrial signals into orbit and back again, the new telescope will help compose a different cosmic cartography, allowing astronomers in Mexico to connect with installations across the world and gather information from radio sources that are light-years away. In November 2024, the site opened its doors to the general public for the first time, to celebrate the "Night of the Stars," a national astronomy festival with telescopes, films, artworks, and conferences. The public, mostly local, was fascinated by the recently opened on-site museum that preserved the original instruments. In a glass case, the museum also exhibited the ground station's logs of significant events, with one

FIGURE 8 Tulancingo 1. Photo by the author.

page open to the register of the 1971 World Cup, and the other to the 1985 earthquake.

Some Final Thoughts

In 2024, in point seventy-nine of her "one-hundred-point plan for transformation," presidential candidate Claudia Sheinbaum announced that, if elected, her administration would oversee the construction of a new Mexican satellite, meant to help close the digital divide between urban and rural areas. Sheinbaum won the election, becoming Mexico's first female president, and has confirmed her interest in using orbital technology characterized as "100 percent made in Mexico" (Rodríguez 2025; in a veiled reference to the controversy around the Morelos system's "nationality") to improve information services. In January 2025, she announced the fusion of the AEM and MEXSAT into a Mexican Space Program (PEM) under the banner of a new institution called the Agency for Digital Transformation and Telecommunications (ATDT). The potential disappearance of the Mexican Space Agency has alarmed many of my interlocutors, as we will see in the next chapter. However, Sheinbaum's focus on communications satellites can be seen as a continuation of Mexican space policy since the 1980s.

Satellites challenge categorization. Both relational and ecological infrastructural assemblages, they are located outside the lower levels of Earth's atmosphere, but they remain trapped in terrestrial orbits by the force of gravity. They are entangled in the planetary, even as they signal the possibility of thinking outside the planet's limits. "Interscalar vehicles," these orbital technospheric objects confuse the categories of the celestial and the terrestrial. They trace irregular spatial, temporal, and conceptual lines between Sputnik and Ulises, Morelos and AzTechSat-1, Mayan astronomy and modern telecommunications, the International Geophysical Year and the Bay of Pigs, babies in Japan and Mexico, Tlatelolco and geostationary orbit, state violence and community defense, cosmopolitan claims and national pride, homemade antennas and technological black boxes, converted telescopes and orbiting debris.

LA NASA NO ES LA RAZA

Ilana Boltvinik and Rodrigo Viñas, of the transdisciplinary art research collective TRES (the public is the third member), have been working with trash in public spaces for years. Their prize-winning works focus on "left behind" objects ranging from plastic bottles filled with urine by travelers and thrown away on the Mexico-Pachuca highway to chewing gum found on the streets of downtown Mexico City and marine debris left on beaches in Hong Kong and Australia. In 2019, they presented *Trapped*, an exhibition about space junk, at Midwest State University in Wichita Falls, Texas.[1] One wall of the exhibition was covered in a printed list of all the tracked objects in orbit. Another wall displayed drawings of some of these objects and a description of the moment when they became space trash:

> November 2007. International Space Station. While repairing a damaged solar array during a spacewalk of more than 7 hours, astronaut Scott Parazynski accidentally lost a set of needle-nose pliers, that floated away and could be spotted drifting below the station.

The exhibition also included photography of the night sky marred by the blurs of fast-moving space trash.

In 2024, TRES followed *Trapped* with another work based on their research into the waste generated by space technology: *La NASA no es*

FIGURE 9 *La NASA no es la raza*, 2024. Courtesy of TRES (Rodrigo Viñas and Ilana Boltvinik). Old media installation, four analog televisions, one parabolic antenna, adapted videos recovered from 1985.

la raza. The exhibition's poster was a copy of an old advertisement for satellite television installation. Piles of cables surrounded the exhibition space. The collective erected a repurposed parabolic satellite dish in the center and placed a vintage television set in each corner. As the antenna revolved, each set displayed a different scene: the collapse of the Televisa building during the 1985 Mexico City earthquake, the launch of the first Morelos satellite, a selection of the television series produced in Mexico that were broadcast during the 1980s, and a series of images about the state of Puebla, where the exhibition was first presented.

After digital cable replaced domestic satellite television service, the antennas that used to signal their owners' entry into orbital modernity became waste, often left in place on Mexico City's rooftops to become "part of the backdrop of technological rubble." For TRES, the "distance between the high-tech aspirations that link Mexico to NASA and the material infrastructure that grounds signals from outer space produces a gap between possibilities and actualities, promises and deceptions, public and private, and so . . . *la NASA no es la raza.*"[2]

CHAPTER 3

THE SPACE GENERATION

So, You Want to Be an Astronaut . . .

I also started to think about the fact that, well, they told me about men who had been to the Moon. And looking at the Moon, and I told myself, "How is it that they got to go, and why don't we all go? I mean, what have they got that we don't, why can't all humans go to the Moon on vacation?" You know, my childish questions.

Carmen Félix, a thirty-five-year-old engineer and entrepreneur, was hopeful. Born in Culiacán, Sinaloa, she was living and working in Amsterdam when we first talked in 2018. She would like to be a "real" astronaut, she told me, but she was working to be a "citizen astronaut," the only option open to her for the time being.

When I was twelve, I started looking . . . it was about how old I was when I got internet access . . . I started looking to see, like, "What are the requirements to be an astronaut?" And obviously what popped up was NASA, right? NASA is for being an astronaut. And there were all their requirements, I wrote down all the requirements. I made a list of the steps, of what I had to do or had to be to become an astronaut. And I remember that at the end of the list, one of the requirements was that you had to be an American citizen. And I said to myself . . . no, well, I can't be reborn, can I? But . . . in the future? The rules will change, or maybe in the future, there will be other . . . available options. And I kept that list, or those steps, always in my mind, you know?[1]

After her parents dissuaded her from studying astronomy by telling her that she "would starve to death, because in Mexico there is no support for science," Carmen decided to study international commerce because she was drawn to the possibility of "learning about different cultures and relating to people all over." But the career bored her, and she decided to accompany a group of young people from Monterrey on a visit to NASA. The trip coincided with the International Astronautical Congress (IAC), so Carmen took advantage of her English-language skills to talk to a few U.S. astronauts, who advised her to study science and engineering if she still wanted to "do something in outer space." Upon her return to Culiacán, Carmen decided that her best option would be to study electronic engineering and telecommunications for its link to satellites. After she finished her undergraduate degree, she applied for a scholarship to study for a master's degree in space sciences in France, which eventually led her to become one of the first Mexican interns at NASA's Ames Research Center, where she worked on projects involving small satellites. The Mexican Space Agency came into being while she was working at Ames, and she participated in its consolidation, taking advantage of her international contacts. She returned to Mexico after her contract with NASA expired and couldn't be renewed because she was a Mexican citizen, but, as the AEM was still in its infancy, she didn't find many work opportunities in her specialty. Therefore, when a contact from her master's program offered her the opportunity to work in the field of space security in the Netherlands, she packed her bags and went back to Europe, where, among other things, she worked as a mentor in the Dutch chapter of the Network of Mexican Talents in the Exterior.

One of Carmen's goals is to involve more young Mexicans in the space industry, and she is an active member of the Space Generation Advisory Council (SGAC), to which we will return below. In our interview, she lamented that, despite the opportunities they have to study in other countries, young Mexicans often return to precarious or nonexistent employment prospects and therefore cannot apply the knowledge they have acquired. Another issue is what has been called the *"maquilización"* of industry in reference to the maquiladora system in which foreign-owned factories in Mexico import raw materials to be assembled into products by an inexpensive labor force and then exported back to their countries of origin. In a similar process, highly qualified recent graduates in Mexico are subcontracted by transnational companies with very good salaries and benefits, but their labor is limited to technical, repetitive activities as they work on a small link in the chain of production. They are rarely allowed to see "the big picture" or contribute significantly in a way commensurate with their skills and knowledge (Nieto 2016).

In what follows, I specifically look at the younger participants in the outer-space milieu that I discussed in the last chapter, members of the Space Generation that have come of age during the era of NewSpace, and for whom Sputnik and Apollo are ancient history. Following Mezzadra and Neilson's (2013) suggestion to use "border as method," I show how the milieu is both inclusive and exclusive, composed of shared experiences, knowledges, and practices, but intersected by a variety of physical, political, economic, linguistic, and generational borders that create a state of *nepantla* in-betweenness.

I contrast the imaginary of a borderless Earth seen from space with the systemic violence that excludes these young people from full participation in the cosmopolitan "space community," a violence that they feel is made bearable by the enchantment of outer space. The restrictions on outer-space access that result from earthly borders become challenges to be creatively addressed through activities such as analog space missions and the use of digital platforms to create temporary milieux simulating borderless spaces, while still being shaped by real-world borders. This chapter looks at the milieu itself as a border zone, an intersection of different forms of knowledge, practice, affect, and identity.

Who Is an Astronaut? Who Is a Mexican?

April 2023. It was the first year that the outer-space sector had been given an entire pavilion at the Mexican Aerospace Fair (FAMEX), held on the grounds of the military base in Santa Lucía on the outskirts of Mexico City, in one of the six buildings constructed to showcase the country's aeronautical industry, civil and military aviation, and the products of its defense technology. Previous editions of the fair only allotted pavilion space for a few stands dedicated to government, university, and private sector outer-space activities. But this year the AEM's area was impressive, including a small auditorium used for lectures, a virtual-reality section, shiny glass cases showing off finely made miniature satellites, and other attractions. A large photo mural surrounded the AEM's stand, including images of a satellite passing over Mexican territory, a man in a white lab coat holding up a CubeSat, a satellite dish from the Tulancingo ground station, two small robots superimposed on the surface of the Moon, the logo of the Mexican Space Agency and two of its campuses, and a group of hands holding up flags from across the Americas. But the stand's entrance was framed by the mural's most arresting image: a collage of enlarged photographs of the "three Mexican astronauts," Rodolfo Neri Vela, José Hernández, and Katya Echazarreta. In the photo, Echazarreta, on the left, is joyous as she exits a Blue Origin vehicle after her space flight; in the middle, Neri Vela stares toward the camera as he floats in zero gravity aboard the space shuttle Atlantis, and on the right Hernández smiles up at the view of Earth from the International Space Station. As I wandered around the space pavilion, I heard a variety of comments about the three figures, ranging from expressions of national pride in these individuals' achievements to criticisms like "He/she isn't really an astronaut," or "He/she isn't really a Mexican."

If we consider the fact that, as many space enthusiasts told me, "Earth is already in space," then we are all astronauts by default. But the distinctions drawn between "astronaut," "space passenger," and "space tourist" continue to be part of a thorny debate over how to classify those people who have been to "space," whose boundaries are more difficult to define than one might imagine. Internationally, the Kármán line, sixty-two miles (one hundred kilometers) above mean sea level, is generally accepted as the border between Earth's atmosphere and outer space. At this altitude, there is no breathable air or scattered light, and, because of the scarcity of oxygen, the sky looks black rather than blue.[2] Neri Vela, Hernández, and Echazarreta crossed the Kármán line and therefore received the astronaut designation. However, for NASA and the U.S. military, space begins at fifty miles (eighty kilometers) above sea level. Virgin Galactic owner Richard Branson reached this height during his flight in 2021, complying with the U.S. definition, although his rival Jeff Bezos, who would make his own trip later that year, quickly pointed out that Branson had not passed the Kárman line.

Ultimately, any attempt to define an absolute boundary is arbitrary, as Earth's atmosphere technically extends to the edge of its highest layer, the exosphere, whose upper edge reaches about 6,000 miles above the planet's surface. However, setting the border of space at this point would mean that the International Space Station, orbiting less than 250 miles up, as well as satellites in low Earth orbits (less than 620 miles), would not be included in "outer space," and, therefore, very few humans (and no Mexicans) technically would be considered astronauts. So, at least while Earth continues to ground human experience, the point at which humans escape its gravitational pull seems to serve a useful definitional purpose. In this sense, all three "Mexican astronauts" underwent an embodied experience that each conceptualized in terms of a transformation in perspective, and that each has leveraged to promote space exploration as a means of improving not just life on Earth but also, more concretely, the well-being of Mexican citizens.

Astronauts, whether space travelers, payload specialists, or flight engineers, are indexes of the cosmic scale: they have touched the stars, and being in their orbit brings us a little closer to our own dreams of outer space. But the cosmic and cosmopolitan dreams of each Mexican space traveler have been traversed by a terrestrial scale marked by borders and immigration, earthquakes, and political relations. And, in any case, in Mexico, as in the rest of Latin America, astronauts are a scarce resource. Space activities in the region do not center on crewed missions, as few Latin American countries have launch capabilities.[3]

In his first, and most popular, YouTube video, Rodolfo Neri Vela (1993) presented a short documentary made eight years after his 1985 flight: *Mexico Calling Atlantis: Documentary of the First and Only Mexican Astronaut*. The video, produced and narrated by Neri Vela, is dedicated to Mexico "and all those Mexican men and women who, in the future, will also have the good fortune to experience such an extraordinary adventure." To many at the time of the flight, it must have seemed that Neri Vela would be the first of many Mexican astronauts. However, this was not to be.[4]

#SoyMexicanoTambien: José Hernández

Retired NASA astronaut José M. Hernández was born in California to migrant workers from Michoacán. As a child, he worked alongside his parents harvesting fruit, traveling back and forth to Mexico each year, and learning English at the age of twelve. Like Neri Vela's autobiography, Hernández's 2012 memoir, *Reaching for the Stars: The Inspiring Story of a Migrant Farmworker Turned Astronaut*, harnesses the poetic power of outer space to inspire generations of young people who "dare to dream."[5] He describes excelling in mathematics as a child, loving *Star Trek*, and watching the Apollo Moon landings on TV, captivated by Neil Armstrong's famous speech about one small step for man: "I had an epiphany. During that exact moment, I discovered what I wanted to be when I grew up: an astronaut or *un astronauta*. And from that moment on, I was determined that absolutely nothing would get in the way of my dream" (Hernández 2012, 35). After hearing about NASA's first Latino astronaut some years later—Franklin Chang Díaz, of dual U.S.–Costa Rican citizenship—Hernández reaffirmed his childhood desire to go to outer space. He obtained undergraduate and graduate degrees in electrical engineering and got a job in a laboratory. In 1992, at the age of twenty-nine, he started researching NASA's astronaut program, listing the requirements in his memoir as a career in a technical field or medicine, five years of experience, and "highly desirable: graduate degree" (134). Unlike Carmen, he did not mention that the need to be a U.S. citizen was a problem. He applied eleven times to NASA before finally getting accepted to its astronaut candidacy program in 2003. He was eventually selected for the STS-128 mission, and, in 2009, flew to the International Space Station on the space shuttle Discovery where he spent almost two weeks as a flight engineer. As a NASA employee, he was not allowed to use the Mexican flag on his uniform; however, he did take a small flag from his parents' country of birth in his luggage, which he later gave to President Felipe Calderón, who had called to congratulate him before the voyage. He also carried five medals of the Virgin of Guadalupe (one for each of his children), a scapular sent to him by a nun from Puebla, and a flag honoring the survivors of the 201st Air Squadron, the only Mexican squadron to fight alongside the U.S. during World War II (206).

As was the custom, NASA controllers played wake-up songs requested
by crew members every morning of the mission, including *"Mi tierra"*
by Gloria Estefan and *"El hijo del Pueblo,"* by Mexican composer Alfredo
Jimenez, the second chosen by Hernández in honor of his parents. Gloria
Estefan had sent an email stating how proud she was that a NASA astronaut was representing the Hispanic community, and even blogger Perez
Hilton commented on his "cool choice" of a wake-up song (Hernández
2012, 220).[6] On the eighth day, Hernández was interviewed by the Mexican press, "the first live interview from space conducted in Spanish"
(219). Again, like Neri Vela almost twenty-five years earlier, Hernández
remarked on the impressive view of Quintana Roo and the Yucatán Peninsula, as well as the invisibility of political borders from orbit. Tortillas
were present, as they had been since Neri Vela's flight. However, Hernández makes no mention of his predecessor in his memoir. He describes
one dinner of soup and steak:

> "What about tortillas? I don't eat anything without tortillas!" I said, jok-
> ingly. "We know, José. You are a true Mexican," responded my commander.
> It is not well known that tortillas are frequently consumed in space. Tor-
> tillas are actually a good substitute for bread; most important, they do not
> produce crumbs that can cause damage to the shuttle's equipment. I was
> so glad I was able to eat something familiar to my taste buds.[7] (218)

As my interlocutors at the AEM had told me, the relationship between
Neri Vela and Hernández has been less than cordial. "They argue about
who is more Mexican and who is more of an astronaut," they laughed,
referencing a conflict that has played out in the Mexican press for years.
In 2009, a reporter told Neri Vela that he, according to Hernández, had
a much easier time going to space because he had the support of the
Mexican government and an agreement with NASA. In response, Neri
Vela speculated that Hernández probably said that he had a more difficult
time getting to space as he had been rejected several times before his
acceptance to the astronaut program. "I don't know how many Hispanics
he had to compete against," said Neri Vela, "but I competed with 800
Mexican-born astronauts" (*"Neri Vela quiere volver al espacio"* 2009).
And in 2018 during Mexico's presidential elections, winning candidate
Andrés Manuel López Obrador thanked Neri Vela for his support during

the campaign, calling him "a great scientist, Mexico's only astronaut." Hernández, who legally claims Mexican citizenship through his parents, replied in a tweet: "With all due respect, Lic. López Obrador, I take exception to your commentaries. According to Mexican law, Dr. Vela is not the 'only' Mexican astronaut. I suggest you read the Mexican Constitution." He concluded his post with the hashtag "*#SoyMexicanoTambien*."[8]

While some tweets posted in reply to Hernández showed support, others included comments like, "As a person and a citizen, you have the right to double nationality and to be Mexican. But as an astronaut, you decided to be *gringo*, not Mexican . . . Until now, the only Mexican astronaut is Dr. Neri Vela," and "According to your logic, Mexico now has many more Olympic medals! Ok, so you're Mexican, but it's a shame you don't show it in the flag that's sewn onto your shoulder."

Space Tourist: Katya Echazarreta

As the first Mexican-born woman to travel to space, Katya Echazarreta has her own Barbie (although it isn't for sale, at least for now). She has been on the cover of the Mexican editions of *Glamour* and *Cosmopolitan*. She has an honorary doctorate for her contributions to society and technology. She co-hosts the YouTube series *Netflix IRL* and is "Electric Kat" on the CBS network show *Mission Unstoppable*. Her website no longer includes blogs written before her voyage: how to obtain internships, how to build a home electronics lab on a budget, how to write a resume, how to start a reading habit, and what holiday gifts to purchase for engineers (headset, keyboard, soldering iron, electrical kit, 3-D printer) and students (iPad, Dr. Martens). But it does list her U.S. and Latin American contact information for those who want to request interviews or speeches, or offer brand deals, sponsorships, and TV casting opportunities, as well as a link to her space foundation's website. Going to space on the Blue Origin mission New Shepard NS-21 in June 2022, the fifth crewed mission for billionaire Jeff Bezos's private space company, has changed her life.

Echazarreta and José Hernández have similar backgrounds. Born in Guadalajara, Jalisco, she moved with her family to San Diego as a child. She studied engineering at UCLA and interned at NASA, where she was later offered a job in the agency's Jet Propulsion Lab, working in a support role on five space missions. Determined to go to space, she decided to participate in a contest organized by the nonprofit organization Space for Humanity (S4H). After a three-year selection process, Echazarreta won a trip off-Earth, beating out 6,999 other candidates from 120 countries. Her flight would be paid for by the sponsoring organization, with contributions from Blue Origin and New Shepard citizen astronauts Cameron and Lane Bess.

The official mission patch for NS-21 announced by Blue Origin looks like an advertisement for a superhero blockbuster. It is shaped like a pyramid, with the last names of the mission participants around the edge. A space capsule emerges from a watery background, an image suggested by Evan Dick, who was making his second trip to space, with the five space travelers below it, black silhouettes with bright blue symbols on their chests, a different symbol for each passenger. Echazarreta's figure features a lightning bolt to signify her interest in electricity and in supporting women and minorities in STEM fields.

The mission patch commissioned for Katya by S4H—also sewn in miniature on the flight suit worn by her Barbie—expresses the organization's mission to "leverage The Overview Effect for peace in our fractured world." Against the outline of a blue sky and nopal cactus, the patch features the silhouette of a dove, a symbol of peace, filled with the image of a starry night sky. Around the edge are the words "Katya Echazarreta-Space for Humanity-Citizen Astronaut #1." The nopal, one of the images featured on the Mexican flag, was chosen for its link to Mexican identity, territory, and cuisine. This "nod to Katya's Mexican heritage" also resonates in a cosmic context, because "this resilient plant surviving across long stretches of time and space with little water or nutrients is much like a space capsule in the desert."[9]

When Echazarreta's mission was announced, users of Twitter in Mexico had mixed reactions. A Secretary of Foreign Relations tweet calling Echazarreta a "source of Mexican pride" was met with a mix of congratulations, questions about whether she should be considered an astronaut or a space tourist, and critiques of the Mexican government for promoting her achievements despite the lack of support from Mexico for her education or her flight. Echazarreta's response to the polemic was "My mother told me something that has stuck with me always: 'You'll never meet someone who is more Mexican than a Mexican who lives outside their country'" (Cortés 2023). She is featured in at least four painted murals around the country. One commemorates two hundred years of diplomatic relations between Mexico and the United States, an initiative sponsored by the U.S. consulate in Tijuana. Another was painted in Cajititlán, Jalisco, the hometown of Echazarreta's grandfather, and an unofficial mural can be found in Atizapán, in Mexico State. The fourth can be found in the "Utopía Libertad," a public cultural, sports, and ecological complex located in the working-class delegation of Iztapalapa, in Mexico City, on the wall surrounding a small planetarium named in Echazarreta's honor.

In August 2023, I was in the audience of a conference given by Rodolfo Neri Vela and sponsored by the AEM in the auditorium located inside the Utopía Libertad, only a few steps away from Echazarreta's planetarium. A mural dedicated to Neri Vela had just been unveiled on a wall outside the complex. In his talk, for which the audience was mainly composed of participants at a summer camp held at the complex, as well as other children accompanied by their parents, the speaker alternated between describing his experiences in space as triumphs for Mexico and lamenting what he perceived as a loss of enthusiasm and support for "the only astronaut who has ever gone to space in representation of the Mexican nation." Times have changed. "Now people only care about 'likes' on social media," he complained. Reflecting on his mortality, he reminded his audience that they "won't have their astronaut forever," and mused on the possibility of a farewell tour of all the Mexican states, like the one he had undertaken in the wake of his 1985 flight on the Atlantis.[10]

"I used to follow Katya on Instagram before she became famous," one young woman told me as she waited outside the auditorium in the very long line for Neri Vela to sign the book she had bought after his talk. "She was amazing. But she changed after she went to space. All this recognition, the fashion shows, the Barbie. . . . I don't think she's sending the right message to young girls. But Neri Vela is the real deal. I get why he's mad."

The Dream of a World Without *Fronteras*

The four wealthy investor crew members of Blue Origin's NS-21 paid for their seats; only Katya Echazarreta and Víctor Correa were sponsored.

Clearly, for the four paying travelers, going to space was an extension of their passion for Earth-bound exploration to an even more extreme context. Correa, who paid a little under $800 for an NFT membership to the Crypto Space Agency (CSA), was chosen randomly among the organization's members to fly to space, taking with him nine "blue chip NFTs," and becoming the world's first "cryptonaut."[11]

On its website, Space for Humanity, which sponsored Echazarreta's flight, describes its mission to democratize access to outer space through sponsoring citizen astronaut space flight and thus creating opportunities for participants "to lead from an Overview Perspective" inspired by the experience of "the Earth as their home and humanity as their community," thus building "a culture of interconnectedness." The overview perspective that serves as Space for Humanity's philosophical basis is derived from the term "overview effect" coined by writer Frank White in the 1980s as a shorthand for the radical change in perspective that some astronauts described experiencing as a result of seeing Earth from outer space. White (2014, 2) defined this overarching concept as a "cognitive shift in awareness":

> It refers to the experience of seeing firsthand the reality that the Earth is in space, a tiny, fragile ball of life, "hanging in the void," shielded and nourished by a paper-thin atmosphere. The experience often transforms astronauts' perspective on the planet and humanity's place in the universe. Some common aspects of it are a feeling of awe for the planet, a profound understanding of the interconnection of all life, and a renewed sense of responsibility for taking care of the environment.

The notion of the overview effect has been enthusiastically received by social actors in the space sector, as it provides a seductive philosophical justification for human activities off-Earth. But White recognized that the "effect's" effects would be limited if they could only be experienced by the minuscule percentage of the human population that has physically traveled to outer space. Space tourism and citizen astronaut programs might increase this percentage, but going to outer space remains out of reach for most people. However, according to White (2014, 67), similar experiences can be produced through technology and communication. This "message from the universe" can be disseminated and replicated around the planet by transmitting still or video images of Earth from outer space, but also using astronauts' public outreach, their bringing awareness to audiences and, hopefully, creating a change in attitudes and practices as more and more humans become, in White's terminology, "terranauts" (70).

Space for Humanity's mission is to convert as many humans as possible into terranauts, making the overview effect "viral" through space outreach. The expansion of space tourism and citizen astronaut programs like S4H has the potential, they say, not just to increase the number of people who have experienced the overview effect firsthand (even though the suborbital view will not be as spectacular as that experienced from the space station), but also to increase the number of people who can leverage this experience into space outreach.

In the video that formed an integral part of her winning application to be sponsored by S4H as a citizen astronaut, and in subsequent interviews in the press, Echazarreta referenced the overview effect in the context of the difficulties she had experienced as an immigrant to the United States. Separated from her family for five years as she awaited her chance to cross the border from Mexico, she recounts, "My Grandmother told me to 'remember that even though we are apart, we are under the same sky.' That message has always stuck with me and speaks to the potential power of the Overview Effect, to understand that we are all connected on the same planet, facing the same challenges, regardless of where we are physically located."[12]

The astronauts that White interviewed spoke of their experiences in outer space in terms of the awe and wonder that the off-Earth visual perspective of the planet inspired. They emphasized the fact that no political borders are visible from orbit, a phrase echoed by Neri Vela and Hernández when they discussed what struck them most about the view from above. Echazarreta also spoke about this change in perspective on geopolitics when talking to the press. Many people ask her what she saw from space:

> But they aren't really asking about the planet, about the atmosphere, the clouds, but "What country did you see? Which states?" And that is something that we have invented, completely. That seems to me to be such an

interesting question, because that's where you can see the differences between my experience of seeing [from space] and their experience of seeing
images or videos of our maps, which is what we know of our world, through
a map. Where everything is separated. "This is the United States, and this
is Mexico, and here is the border, and you can't cross it unless you have
permission." But you see everything from above, and none of that is visible!
It's all desert, all mountains, all clouds, those lines don't exist. (Platzi 2023)

White promotes his terranauts as agents of radical change that work
"toward a sustainable and peaceful planet." Yet he goes even further in
claiming that the final goal of human spaceflight is not just the creation
of a healthy relationship between humans and Earth using the overview
effect, but the transformation of humanity into a spacefaring and space-
inhabiting species as part of an evolutionary process that continues what
he sees as the innate human drive for exploration and expansion, a vision
that aims to establish "a universal civilization, a golden age, humanity
taking its rightful place as citizens of the universe" (White 2014, 172). In
a later work, he makes the grandiose declaration that space exploration's
aim is not simply to benefit humanity, but to "spread life and intelligence
throughout our solar system and beyond," making the universe "more
alive, intelligent, and self-aware," and contributing to cosmic evolution
(White 2019, 25).

But as Marianne Janack warns (2002, 273), this "God's eye trick" has
the effect of allowing "the views of some-people-in-particular" to pass
as "the view of no-one-in-particular." In the words of Vandana Singh
(2023, 97),

We live in a Eurocentric globalized culture, which means that a one-world
conception arising from those who are part of the dominant group may
gloss over and erase concepts, dreams, and imaginings of other humans. It
is also important to remember that patriarchs, feudal lords, and colonizers
have always had a rather different interpretation of a vantage point from
which they can see all they own, lords of all they survey. Here we may well
see, not so much a sincere acknowledgment of the unity of life on Earth,
but instead, a triumphalist validation of the power of the powerful, and
perhaps an exacerbated desire for control and dominion.

In a similar vein, Jordan Bimm (2014, 42) argues that the overview effect should be seen as the product of a cybernetic view of the planet, linked to the paradigm of systems thinking that emerged after the Cold War. And, for Ceridwen Dovey (2020), romantic views of humanity in space, looking benevolently back toward Earth, "ignore the fact that space is already a militarized and commercialized zone." Dovey goes on to write about the very citizen astronaut contest sponsored by Space for Humanity won by Echazarreta:

> One might want to read the small print carefully. While the organization is happy to bankroll your experience of the overview effect, it is not prepared to permit you the freedom to express your epiphany in whatever way you see fit: returning to Earth as a newly charged eco-warrior, for example, campaigning to protect Earth orbits from being further trashed by tens of thousands of small satellites being launched by private companies such as Musk's SpaceX; or feeling moved to dedicate your life to efforts to turn the Moon into a protected natural and scientific reserve, free from commercial mining.

She proceeds to call attention to a small caveat buried at the bottom of the Space for Humanity website: selected citizen astronauts "must also be willing advocates for the growth of the commercial space industry." Dovey concludes, "No matter what you experience up there, the organization doesn't want to hear about it unless you come back down and join them in their celebration of the art of making money in space."

To be fair, Echazarreta's position is complicated. She flew aboard a mission launched by a predatory transnational company and was sponsored by an organization that professes a belief in an unproblematized one-world dream. However, although she shares both Bezos's and S4H's faith in technology and progress, her interest in the overview effect is centered on the resolution of conflicts and inequality in Mexico rather than simply the transmission of a whole-Earth perspective or the advancement of cosmic evolution. Neither does she present herself as a universal subject. Her standpoint at least partially shifts the universalist discourse of White and his like-minded collaborators to a more grounded recognition of earthly difference. Many of the astronauts interviewed by White marvel at the way that borders disappear when Earth is viewed from outer space. But as a female Mexican-born immigrant to the United States, Echazarreta has been transected by borders in ways that most space travelers have not. "I'm a woman, I'm Mexican, I'm dark-skinned, I'm short—I'm one of the shortest people who's gone on a space mission—if one of those things by itself is enough to experience discrimination, imagine what happens when all of these things that I am are combined. So it has been hard" (Cortés 2023). Although she highlights the potential of the overview effect in her press interviews, much of her work revolves around promoting Mexican space activities and increasing opportunities for Mexican women to study and work in STEM-related fields.

In his memoir, José Hernández espoused the values of hard work and persistence as a way of overcoming adversity. But Echazarreta, while acknowledging the value of having figures that inspire people to go beyond societal expectations and limitations, does not gloss over the difficulties faced by women and minorities.

> Men still think women are there to serve and can't accept that they have other choices. . . . Girls are still given toys related to cooking, cleaning, taking care of babies, and boys are given Legos, toy cars, engineering kits, building blocks, toys that let them develop mentally, but girls are told "You're going to take care of babies or clean the house." (Cortés 2023)

Conscious of the difficulties faced by young people who want to pursue space careers in Mexico, she laments, "How do you tell a little girl who wants to be an astronaut that she'll have to leave her country if she wants to fulfill her dreams?" (Platzi 2023). She created a foundation to develop technoscientific opportunities in Mexico and lobbied the Mexican government to provide more support for outer space; the efforts of Echazarreta, the AEM, and a handful of politicians and scientists, have, at the time of this writing, led to a possible constitutional reform that recognizes outer-space activities as priorities for the nation, a bid for national technological sovereignty and investment in science and technology education that will, according to Echazarreta, ensure that "Mexico's future will be for Mexicans" (Cortés 2023).

Notably, her view is not always from above. In one interview, after extolling the overview effect and the lack of visible borders from space, she mused:

You look up and you see a cloud here in Mexico City. And well, it's a cloud. But you see it from up here, and it's a cloud that's connected to a system of clouds in Puebla, and another system of clouds in Morelos. So someone in Morelos is seeing this cloud, and we're seeing this cloud, and we don't realize that really, we're connected. (Platzi 2023)

This echoes the comments made by the telecommunications engineers at Eutelsat about garbage being thrown away in Monterrey affecting Mexico City. These references to the atmosphere as part of a planetary system perform a double function: on the one hand, they acknowledge Earth as an interconnected hyperobject, what Timothy Morton (2013, 2) refers to as a "thing that is massively distributed in time and space relative to humans." But, on the other hand, they insist on planetary situatedness and particularity. The overview effect assumes a global view (which is always partial—it is impossible to see all sides of a sphere at once), but Echazarreta and the Eutelsat engineers are looking at Mexico. Looking up is a form of worlding, a way of comprehending and living on Earth at an inhabitable scale.

Going to Space Without Going to Space

One of the Echazarreta Foundation's projects is the establishment of summer space camps for Mexican children with few economic resources. The last part of each session is dedicated to a miniature analog mission that allows the children to have an experience "as if" they were in space, complete with space suits and a lunar or Martian landscape. Like Carmen Félix, Echazarreta had been involved in some of these analogs long before she was chosen for the Blue Origin mission, as a member of "Project PoSSUM," now called "PoSSUM 13," a program created to support students "especially young women—who have a passion for space science and exploration" in part through participation in analog experiences.[13]

Analog missions occur in "extreme" sites like Antarctica's tundra and deserts in Chile and Utah, or habitats and laboratories all over the world, allowing scientists to test objects, processes, and human bodies to visualize and address potential problems in a lower-risk (and lower-cost) environment than outer space. All astronauts undergo some kind of analog or simulation training when preparing for space missions. But analog missions also provide simulated space experiences for "citizen astronauts," who are unlikely to physically leave Earth. In Latin America, they have become an important way to engage with space science and technology, promote the development of national and regional "space ecosystems," and motivate young people to study for careers in STEM fields, within the region's infrastructural and budgetary constraints. It is perhaps not surprising that the topic of analog missions was one of the first to come up in my initial research with the Mexican Space Agency, given that, as a country that does not have a crewed space program, "as-if" projects are perceived as an important medium for Mexico's involvement in the space sector.

Mexico's first analog experience came about when Apollo astronauts participated in training missions in the Pinacate desert in Sonora between 1965 and 1970. Craters called "Cerro Colorado," "El Verdugo," "Molina," and "El Elegante" stood in for the lunar Fra Mauro. In the depths of El Elegante, a rock still bears a trace of one NASA crew in the form of graffiti: "Apollo14, NASA 2/16/1970" (Arreola 2017). The next analog missions took place after a watershed event in Mexico's space history: the 2016 IAC, which was held in Guadalajara. As I noted in a previous chapter, well-known astrocapitalist Elon Musk gave the keynote speech, "Making Humans a Multiplanetary Species," in which he invoked the planet Mars as "possible . . . something that we can do within our lifetimes."[14]

Musk argued that humanity can follow one of two paths: stay on Earth and, one day, face extinction, or leave Earth and become a space civilization and multiplanetary species. The second path, Musk insisted, is the only way to survive. "I hope you all agree with me, yes?" The audience yelled and clapped. The image of Mars appeared behind him, a red sphere filling the screen. "This is what we want." More cries and applause. "You are the Mars Generation," he told the students in the audience. "You are the future." He did provide some nuance to these inspirational slogans when a young woman in the audience asked him when he would hire international workers at SpaceX. "We aren't allowed to do that," he replied, referring to the International Traffic in Arms Regulations (ITAR), which are infamous among Latin American astronautics students and make it difficult for non-U.S. citizens to work on potentially sensitive technological projects. "But," he exclaimed, in an appeal to the maquiladora system I mentioned above, "Tesla will hire you!"[15] Almost everyone

I talked to agreed that the event was a turning point for many young people in Mexico. "It was a moment of effervescence," Juan Carlos and César recalled, "there were a lot of people doing a lot of things, from all over the country."[16]

In December 2018, with Carmen Félix's support, MEX-1, the first all-Mexican analog team, was accepted for a mission at the Mars Desert Research Station (MDRS). I spoke to some of the team members on Skype a week before their trip. All were members of the SGAC, and they had met at the IAC in Guadalajara. They were supposed to go to the MDRS in November 2017, but between budgetary problems and the aftermath of the 7.1 magnitude earthquake that had shaken Mexico two months earlier the mission was postponed for a year. The members of MEX-1 lamented the lack of opportunities in the field of science and technology in Mexico, especially in certain regions. They emphasized the importance of investing in those fields "for the good of the country's future" despite the political limitations of government support. Through the AEM, they argued, the Mexican government should provide more backing for outer space so that highly qualified young people don't have to be "citizen scientists" or look for professional opportunities as isolated people outside their country. "Innovation generates progress." These young engineers, who had not been born when Neil Armstrong landed on the Moon or Neri Vela went into space, echoed Carmen and Katya's dream: "It's really hard right now, but why not pave the way so that Mexican astronauts can once again be included in [space] missions?"[17]

"I would have loved to see people doing what we're doing now when I was a child," César told me. "And I don't want to be a governor or the president of the Republic. You can make change in a different way, without being in the spotlight." Guerrero, his home state, is plagued with narcotraffic and kidnappings, he said, and he had friends who been victims of violence for being in the wrong place at the wrong time. "It makes me furious, and it irritates me because it's so hard to 'change the chip.' If you want to see what Guerrero could be, look at Neri Vela, look at me." His dream for the future is that technological objects stop being "made in Mexico" and start being "designed in Mexico." He wants Mexico to become "a country that can generate value for the world."

Tania agreed. "Mexicans are always looking for a hero, a role model. I want Mexicans to stop saying 'I want to be El Chapo,' but rather 'I want to be an astronaut." I want children and young people to know that in Mexico you can do things, you can be an astronaut. That they feel inspired by space, just like us." Genaro listed some of the things that space technology can accomplish, including communication, progress, local economic growth, disaster management, and support for displaced people. "Young people can take up the standard so that we aren't a country of technological illiterates. Space is a generator of solutions, and it's not a pipe dream. It doesn't mean colonizing other planets, or *Star Wars*, or *Star Trek*."

MEX-1 would be Tania's second analog mission; she had already participated at the MDRS as a crew member of the second LATAM team in 2018. I asked her if she had felt like she was really on Mars. The most "authentic" aspects of the experience, she said, "were not being in contact with anyone, being closed-in, with no time alone." She continued, "I started to feel mentally exhausted during the second week. You fight with your teammates, and little things get big. I started to miss the color of an apple." But ultimately, "how real it is depends on how seriously you take it." And how did it feel to return to "Earth"? I asked. She replied, "If you go to space, you are going to feel alone, abandoned. But when you're alone, it makes you think about being with humanity, being connected. At the same time, going back to all that information, to Earth, is overwhelming." She was curious about what her new teammates were expecting from the experience. Genaro replied that he wanted to experience total darkness and complete silence, to "share the universe with the universe itself."

National media outlets sometimes blur the distinction between astronauts and analog astronauts. Yair Piña, one of the first Mexican students to participate in an analog mission after Carmen Félix, made headlines in newspapers and on Twitter for being "invited by NASA to travel to Mars," although later articles clarified that he would be part of a Latin American analog mission at the MDRS, in a project that was not affiliated with NASA (Mulato 2017). He was invited to meet then-President Enrique Peña Nieto at his residence in Mexico. Some of my early interlocutors at the AEM joked (?) that they were unsure whether Peña understood that Piña was not actually going to Mars. Piña echoed Neil Armstrong's famous words in a press statement, clearly referencing one of then-President Donald Trump's border control strategies: "It's a small step for me, but a giant step for Mexico toward what we seek for space

FIGURE 10 Genaro Grajeda, MEX-1 Satellite Officer / Health and Safety / Journalist. Courtesy of César Augusto Serrano.

exploration. Walls can be built, but we are committed to building bridges for peace and love."

One interlocutor who does "serious space missions" told me he thought analog missions were just "playing at space," and that in their attempts to highlight Mexican accomplishments the press were contributing to an overvaluing of activities that do not make a real difference in the advancement of the space industry.[18] But Oscar Ojeda, a Colombian astronautical engineer I met on Twitter, does not agree that analog missions have no value. He distinguishes "astronauts" who train to go to space and "crewmembers of analog missions." He was also at the 2016 IAC and SGAC meeting in Guadalajara, where, after studying the more technical aspects of space engineering, he "came face-to-face with the world of human spaceflight."[19] Along with Tania, Oscar participated in the second Latin American analog mission at the MDRS in 2018, the site's first all-Colombian mission in 2019, and a lunar analog mission at the HI-SEAS research station in Hawai'i. His experiences motivated him to create an analog habitat in his native Colombia, to "build a pipeline so that the knowledge we generate here doesn't stay here but can be used by top researchers in other places." He and some colleagues won a small grant from an SGAC call for financing space initiatives. "Like a Kickstarter," he said. This seed money helped him start the Cydonia Foundation (after the region on Mars famous for the feature known as the "Face on Mars") and build a habitat outside Bogotá, in the municipality of Chía ("Moon" in the Muisca language).

Cydonia built the HADEES-C habitat at a low cost, with an open-source design so that it can be replicated. Everything was bought at Home Depot or a local hardware store, "to show what you can do even in countries that don't have space agencies," and all the builders were volunteers.[20] Eventually, missions will produce scientific research and technological innovation, but other objectives, like fostering creativity and imagination, are also important, Oscar told me, comparing these "open" missions to role-playing games that allow players to speculate about possible worlds. He is explicit about his goal: to make space accessible for people in Latin America and "to decolonize outer space and the outer space industry."

Originally, the "citizen" of "citizen science" was imagined as belonging to a civic category, implying membership in a universal civil society. But, in the context of Latin America, it also points to the existence and negotiation of political borders, even in the cosmopolitan outer-space industry. As political philosopher Étienne Balibar (2016, 275) points out, one is never simply a citizen, but always a citizen of somewhere. For Carmen and the members of MEX-1, being citizen scientists implies a paradox: even as their education includes them in a global community of space professionals from which others are excluded, the political limits of citizenship are what prohibit them from being "real" astronauts. They travel on legal visas, study at accredited universities, obtain international grants and scholarships, and speak English. They are "cosmopolitan." But their experience is also necessarily cosmopolitical, transected by borders, identities, and impossible Silicon Valley–style aspirations.

For this reason, the worlding aspects of analog missions in Latin America are key, as they imagine not only life off-Earth, but also a world in which there is equal access to earthly and spatial resources and where, truly, "space is for all." Lisa Messeri (2016, 30) writes about the analog missions undertaken at Utah's MDRS as providing a "double exposure," a superimposed temporal and spatial image of a Martian future with a terrestrial present (see also Olson 2018, 41). In Latin American contexts, where few countries have launch capabilities or crewed space programs, analog missions provide a "triple exposure": a physical present on Earth, an imagined future on Mars, and another imagined future on Earth in which Latin Americans are equal participants in the global space milieu.

The Space Community

The young participants in the space industry I've talked to are passionate about a future in outer space but more ambivalent about the possibilities of studying or finding work in the field in Mexico. Their critiques tend to center on four sources of friction: discrimination they face outside Mexico, a national culture that undervalues innovation and collaboration, a lack of governmental support for science and technology, and a generation gap between "old space" and "new space" actors. "Old space" actors may remember Cabo Tuna and Apollo, and they may have worked on the Morelos satellites or similar projects in the 1980s and 1990s. They have comfortable jobs in government agencies or established companies and, some say, still set Mexico's space agenda. They are "boomers" and sometimes "old Gen X." "New space" actors are "millennials" and younger, born long after Apollo and Neri Vela's flight. They came of age after the end of the Space Shuttle era and were inspired by the dawn of the NewSpace Age. They have been promised that space is for everyone but often feel trapped between that promise and a reality that keeps them on the ground. If the "old space milieu" was centered on national institutions and the development of a traditional aerospace industry, the "new space milieu" with which it overlaps has been shaped by digital platforms, transnational networks, and entrepreneurial aspirations.

Tania was in Europe, so I contacted Juan Carlos and César to find out how the MEX-1 analog mission had gone. We had coffee in an old colonial house in the upscale Condesa neighborhood of Mexico City that had been converted into a co-working space owned by Google. Their company Dereum Labs had team meetings here and, thanks to a contact in the AEM, Google did not charge them for the use of the space, which was a kind of border zone between Mexican business culture and global tech culture. A quote in English by Eleanor Roosevelt was pinned on the door of the women's bathroom: "The future belongs to those who believe in the beauty of their dreams."

Juan Carlos, César, and eight other partners co-founded Dereum in 2017. After a few conflicts with some of the other co-founders over how to divide future profits, the company was consolidated around engineers Juan Carlos and César, who were later joined by industrial designer Kaori and an "angel investor." They design rovers ("everyone was doing CubeSats; we wanted to make a CubeRover") for lunar exploration, and various transnational companies have shown interest in their project. (A couple of months later, I was invited to a pitch cocktail party for Dereum at Mexico City's Dow Industries headquarters, and I would see them again in 2022 at the FAMEX presenting a prototype of their Jaguar rover—the name was the suggestion of a marketing company—at the Airbus stand.) Along with robotic design and engineering, they have recently begun to help other companies write business models that incorporate expansion into outer space.

I asked why they thought the space industry wasn't advancing, even after the 2016 IAC. The positive energy didn't last long, they lamented. "The effervescence faded because there wasn't a lot of willingness to collaborate and work as a team." There is a lot of *protagonismo* in Mexico, and "everyone wants to be first." I asked whether they thought there was a Mexican space community. Juan Carlos thought there is, "because people are passionate about space." The problem, he said, is the lack of a "space ecosystem" to support space activities. César agreed that there is no space ecosystem but also argued that there is no space community. "Because people don't want to collaborate." His community is more intimate: his company, his colleagues, university clubs, and groups on Facebook.

Other young people also question the existence of a Mexican space community. After MEX-1, Tania got a job at the AEM, but she was soon disillusioned with what she perceived as a lack of meaningful projects and a male-dominated work culture. She is currently studying space science at the International Space University in France. Kaori became burned out after working at Dereum for several years. She left the company and has transferred her passion to starting one that uses space to inspire others to imagine alternative futures. And before Oscar created his foundation, he confronted discrimination in his astronautics master's program in the

U.S., where his advisor (thinking his Colombian student was Spanish, I suppose) told him he would be "better off cooking tapas and paella." He also faced obstacles when applying for coveted NASA internships, thanks to the ITAR restrictions mentioned by Elon Musk in Guadalajara.

Kevin, an undergraduate engineering student from the UNAM, participated in several analog missions in the U.S. and Europe. However, he lamented that as analog missions become more common media interest and institutional support have waned. On his first mission, he received a lot of press coverage. It seems that, for Kevin, analog missions only create temporary communities, as the feeling of support and camaraderie was fleeting. He criticized those he called "astrofriends," who only used him for contacts and information but disappeared when he needed emotional support. "You can't see borders from space, but here, it's all borders. The dream is over, and now it's time to wake up."[21]

The Space Generation

Since the 2016 IAC, the SGAC, a nongovernmental organization created in 1999, has attracted many young Mexicans to its ranks, drawn to the promise of belonging to this aspirational, transnational space community, although when I talked to Juan Carlos and César it seemed that the SGAC was going through a lull in Mexico. One of the SGAC's other members told me they thought that, apart from the "old space–new space" divide, there was also a breach between the millennials who had been active since 2016 and the Gen Z space influencers who "seemed only to be interested in taking space selfies." However, in the past year or so, the SGAC has been reactivated by young people in STEM fields, as well as by students and young professionals in the social sciences, particularly the field of outer-space law.

According to its vision statement, Space Generation is "THE global network for students and young professionals interested in the space industry." It exists "to employ the creativity and vigor of youth in advancing humanity through the peaceful uses of space" and to "nurture the next generation of space leaders." It has Permanent Observer status in the Committee for the Peaceful Uses of Outer Space of the United Nations (COPUOS) and has agreements with "leading space agencies, corporations, and organizations from around the world," working in tandem with the International Space University (ISU) and the International Astronautical Foundation (IAF). The SGAC is organized into six regional sections, each with two elected regional coordinators, and is governed by an international executive committee. Each of the 165 member countries has one or two "National Points of Contact" (POCs), and current membership is estimated at around twenty-nine thousand.[22] Following United Nations regional divisions, Mexico participates in the "North, Central America, and the Caribbean" section of the SGAC, which the organization's web page refers to as "one of the most diverse regions of SGAC." Membership is limited to those thirty-five and under, which makes the organization youthful and dynamic. "It's a good place to try things out," says Oscar. "However, it can be difficult for STEM students in Latin America to finish school and get competitive internships before they turn thirty-five," he continues. The SGAC tries to be global, but local and regional realities intrude.[23]

According to conversations I've had with SGAC members in Mexico, this division has both benefits and drawbacks. On the one hand, Mexico develops relationships and shares resources with its wealthier northern neighbors. On the other hand, it does not always participate equally, given the lack of support from the Mexican government and the difficulty of finding sponsors for events and activities.[24] Since it does not participate in South American events, Mexican students and young professionals in the space field must also find other avenues for the promotion of Latin American collaborations.[25] The 2023 regional meeting of the SGAC was held in Costa Rica. A group of attendees from Mexico met at the meeting and decided to revive the SGAC in Mexico by organizing the country's first national event, to be held in the town of Tonantzintla, in the municipality of San Andrés Cholula, Puebla.

As we saw in the first chapter, the chapel in Santa María Tonantzintla is a fascinating example of baroque *nepantlismo*, with its native cherubs brandishing bows and arrows. It is also the site of both the UNAM's National Astronomical Observatory (OAN) and the federal National Institute of Astronomy, Optics, and Electronics (INAOE). The SGAC event's logo, printed on keychains and coffee mugs, is a miniature cosmograph. Its central image is of the volcano Popocatépetl, the "smoking mountain." The volcano's highest point, along with that of his partner the "white woman" Iztaccihuatl, is home to a few of the last agonizing meters of glacial ice in Mexico, and both peaks are unmistakable features of Tonantzintla's landscape. In the foreground of the logo is an outline of the Chapel of Santa María. Do the shadows in the base suggest something older buried beneath the colonial structure? The backdrop is a starry night sky in shades of blue and purple.

The event was inaugurated by one of the INAOE's directors, who reminded the audience that "Tonantzintla and the Cholulas are millennial cities," and that, again, "we have always looked to the stars." After listing the important people present and briefly describing the event, he concluded, "Mexico's youth is ready for the challenge of outer space, and the sky is no longer the limit."[26] The audience was primarily composed of "Mexico's youth," the Mars Generation or the Space Generation, who would spend two days in Tonantzintla learning about opportunities in the space field, discussing its challenges, and, perhaps most importantly, networking. There were representatives of the AEM, the incipient Latin American and Caribbean Space Agency (ALCE), and the Federal Institute of Telecommunications (IFT), as well as universities and private companies. Quite a few women were in attendance and in the leadership of the SGAC's organizing team. There were fewer women representing institutions on the panels, however. During coffee breaks, WhatsApp QR codes and LinkedIn contacts were exchanged.

Talks and roundtables included "Advances in Telecommunications in Mexico," "The Space 2030 Agenda and Space Governance," "The AzTech-Sat constellation," "Space Education," "The NFT Luna Mission," and "Dark Skies." Most of the discourse echoed what I had heard at countless other space gatherings in Latin America: "NewSpace changes everything because it democratizes outer space"; "For Latin America, regional alliances and teamwork are critical, especially because we don't have a decent budget"; "We need more investment from the private sector"; "If we develop our own satellite technology, we will be technologically autonomous and won't have to buy information from other countries"; "Investing in space technology will close the digital divide"; "There is room for all disciplines in space. Soft skills are important, too." To the question "Who owns outer space?", someone calls out, "No one!" but she is ignored. "Humanity! And Mexico is part of humanity. Our outer space is waiting for us. Our future is in outer space. Humanity's future, and Mexico's future."[27]

These presentations were interspersed with students' poster and paper presentations, and group work on what the organizers considered to be priority topics for this year: Space Robotics, Rocketry, Space Entrepreneurship, Astrophysics and Planetary Sciences, and Space Security and Conflict Prevention. Christopher, a philosopher from the UNAM, and I had been assigned to guide this last working group, which we retitled "Space Ethics." Each group, oriented by "experts," was asked to "workshop" a series of recommendations for the international SGAC.

In our group, we presented a series of cases, both real and speculative, intended to create a series of ethical principles that the group felt should guide human activities in outer space. The members of one rocketry team had signed up for this group—I remembered them from the ENMICE in Chihuahua. They confessed to having been a bit blindsided by the new requirement to include a social impact analysis in their project, which is why they decided to join this group instead of the rocketry group. Although they wanted to keep participating in the ENMICE, their goals were more ambitious, as they had set their sights on international rocketry competitions like Spaceport America. Travel to New Mexico where the competition is held is complicated, though. They've had to rely on contacts with Mexican politicians to get visas.

During the last session of the event, the working groups presented their conclusions. I admit that I got a little choked up when my group presented their "Code of Ethics for Outer Space Activity" emphasizing social and environmental responsibility and arguing for the principles of diversity and inclusion in the decision-making process. The event concluded with the awarding of prizes to the poster presentations, whose authors have the chance to publish their works in a NASA repository, and the presentation of the awards for "emerging space leaders." The event was officially closed by another of the INAOE's authorities: "I will conclude with the words of Elon Musk: 'The future belongs to those who believe in their dreams.'" A few weeks later, I would attend another space event and hear Itzel, a young space lawyer and one of the organizers of the SGAC meeting, complain: "I am tired of hearing people say that if you can dream it, you can do it. It's not enough to dream; you need hard work. And sometimes even that's not enough."[28]

The event was centered around professional development, but it also tried to fortify the "Mexican space community." Networking was key, but it was not the only strategy for community building; for example, the event program included social gatherings such as "Mexican night," which was held at a local bar after the first day's sessions. Attendees were supposed to dress in "typical attire," although only a couple of people complied. There were tacos, drinks, and karaoke.

Alongside the physical interactions, a Mexican design company created a virtual meeting space—an analog of sorts—in which participants' avatars could interact, upload material from the working groups, and roam around a digital space divided into rooms. The "cockpit" or main page was an orrery, and avatars could travel from there to a gallery of images, a virtual model of the Space Shuttle, an auditorium where an astronaut in a space suit waved from the stage—which also featured a model rocket and a NASA logo on the screen—or one of the portals designated for each working group. Each portal had its charm, but the one designed for the space ethics group was, not surprisingly, my favorite. It was designed as a campsite next to a lake, complete with the sounds of chirping crickets and crackling logs. In front of the water, the United Nations and Mexican flags waved, flanking a collage of images from our sessions. A woman in a space suit knelt on the ground in front of a spinning black hole. A rocket flew over the half-sphere of an Earth embedded in the ground. A model of Sputnik floated above the scene, accompanied by images about Yuri Gagarin's life. The Moon and a model of the ISS hovered against the backdrop of an aurora, and an optimistically balanced scale perched on a rock next to the campfire.

The platform was designed to provide a space for participants to continue interacting after the event, which was fortunate, since the overtaxed internet at the INAOE did not permit the full experience. I visited it again as I was writing this chapter, a couple of months after the encounter in Tonantzintla, but no one else was around. So, I sat my avatar down on a log and listened to the crickets.

Some Final Reflections

In many ways, members of the Mexican space milieu, particularly young people, are nepantla, in León-Portilla's sense of being caught between two worlds rather than the philosophical Nahua sense of interaction from the middle. In my relations with young people in the space milieu over the last few years, I've found myself infected with their moods. When they are depressed about the lack of opportunities in Mexico and promises unfulfilled, I get depressed, too. But when they discover or rediscover the wonder of outer space, I am thrilled along with them, even when I don't necessarily share their dreams for a space-faring future. My interlocutors have delighted me with stories of encountering the cosmos on a starry night and through the writing of Bradbury, Asimov, or Sagan; imagining aliens; writing a fan letter to Rodolfo Neri Vela and getting a response; getting an internship at NASA; programming a satellite or winning a rocketry competition. Their relation to outer space is one of hope and enchantment, an effect that Jane Bennett (2001, 3) describes as "a mood of fullness, plenitude, or liveliness, a sense of having had one's nerves or circulation or concentration powers turned up or recharged—a shot in the arm, a fleeting return to childlike excitement about life." And then they recount narratives of discrimination and exclusion, disillusionment with space institutions, and lost opportunities.

The "final frontier" imagined by U.S. space enthusiasts hits differently in Mexico, in part because here there is no historical imaginary populated by pioneers traveling ever further westward, and in part because the word *frontera* translates as both "frontier" and "border." Despite universalist pretensions, the world imagined by the overview effect is only an apparition, while the world created by the refrain (Stewart 2010) "space is for everyone" is built on exclusions. My interlocutors are pierced by borders: between space and Earth, Mexico and the United States, the North American / Central American / Caribbean SGAC and the South American SGAC, OldSpace and NewSpace, childhood and adulthood. These borders produce nepantlismo, mostly the kind decried by León-Portilla (1962). Young participants in the Mexican space milieu are nepantla for their generational status, nepantla for their migratory status, and nepantla for moving between their "spacefaring" aspirations and "emerging nations" reality. But, as Mezzadra and Neilson (2013, 284) argue, borders can both connect and disconnect, "constructing provisional and contingent unities that are always in negotiation and context-specific." So, being crossed by borders also produces the other kind of nepantla, the kind that Anzaldúa (2009, 310) describes as a way of living in/on/with the border that is also a bridge, a "space between the way things had been and an unknown future." And borders are also milieux, where different forms of knowledge, practices, identities, and worlding potentials intersect. Yet the worlds generated, those that are being brought into being through decolonizing analog missions, space camps, and SGAC karaoke, are precarious. They depend on young people's continued passion and hope for the future, which in turn rely on structures of support that, so far, are unreliable, nonexistent, or openly hostile.

As I write this chapter in early 2025, however, there may be a new communal impulse on the horizon. The uncertainty surrounding the future of the Mexican Space Agency has motivated actors in the space milieu(x), particularly young people, to create platforms for the exchange of opinions and proposals for Mexico's future(s) in outer space. WhatsApp groups, physical meetups at bars in Mexico City, Zoom meetings for geographically dispersed participants, and collaborative documents have generated spaces for the formulation of a series of collective responses to the disappearance of the AEM. As a (mostly silent) member of at least four different chat groups, I have read many texts criticizing the beleaguered space agency for its miniscule budget (all concur), lack of vision (many concur), accusations of corruption (some concur), and failure to support young people's educational and professional development in the space sector. The last complaint varies according to peoples' personal experiences. Some interlocutors in these spaces have celebrated the demise of the AEM for its perceived failures, but most have attempted to formulate alternative proposals that would fortify the agency rather than destroy it, arguing for the importance of an internationally recognized governmental institution dedicated to the integration of outer-space activities. In some of the links to social media circulated in these groups, Rodolfo Neri Vela has called the fusion of the AEM with MEXSAT a "major setback for the space industry,"[29] while Katya Echazarreta has been more circumspect.[30] It remains to be seen if this sense of crisis will forge a greater sense of community in the Mexican outer-space milieu, or if it will further increase the precarity of the Space Generation's hoped-for futures.

MATTERS OF GRAVITY

April 2024. I had been invited to a small dinner party at contemporary artist Ale de la Puente's house, commemorating the tenth anniversary of the project *Matters of Gravity*.[1] Ale and Nahum, director of the KOS-MICA Institute, produced and curated the project, while theoretical physicist Miguel Alcubierre, also a guest at the party, participated as the science advisor.[2] Other artists included Juan José Díaz Infante, Marcela Armas, Arcángelo Constantini, Iván Puig, Tania Candiani, Fabiola Torres-Alzaga, and Gilberto Esparza.

As we ate vegetarian quesadillas and drank sparkling wine, Ale, Nahum, and Miguel talked about the impact *Matters of Gravity* had on their lives. None of them would forget the experience of traveling to Star City in Russia to make art in a zero-gravity environment. There is nothing like the sublime feeling of floating with nothing to hold you in place. Of course, achieving the sublime can be a deeply unsublime process. The team underwent both theoretical training (Miguel gave lectures on gravity and astrophysics) and physical training in Mexico ("practice" on the extreme rides at Mexico City's Six Flags) and in simulators in Russia, as well as medical tests of all kinds. Financing was complicated—should they look for grants for art or science? And the project was almost canceled over geopolitical tensions with Russia.

Finally, after two years of planning and preparation, the eight artists and one scientist experienced forty different changes in gravity during the ten parabolas made by the gargantuan Ilyushin 76 MDK aircraft, originally used as a tank transporter. There was normal gravity, then double gravity, then thirty seconds of zero gravity, then double, and normal again during each parabola. "Everyone was throwing up," laughed Nahum.

The artists created twelve artworks, which would be exhibited in Mexico City, Moscow, Slovenia, and Texas. Among them were Ale's piece . . . *And Earth is in Zero Gravity*, for which she released a globe during the flight: "Abandoned to forces (both cosmonauts and artists), the globe revolved in expected and unexpected ways, materializing mythology and creating an image of the invisible fate of our planet" (De la Puente et al. 2014). Marcela Armas's work *Geopsy* combined outer-spatial matters of gravity with Earth's geological history: she brought a geological sample from northern Mexico onto the plane where it lost its gravitational force, serving as a prompt for formulating "questions about the relative weight of things and the possibility of thinking and believing that other forms of writing history are conceivable, but especially, of allowing ourselves to face historical events" (De la Puente et al. 2014). Given the immense logistical and financial challenges of the project, Nahum "wanted everyone to stop focusing on switching on the camera or doing their own experiment and start interacting with each other." His piece was titled *Holding Air* and revolved around a series of embraces between the participants in zero gravity, attempts to make contact and exchange affection in conditions that make an embrace "impossible" (De la Puente et al. 2014).

And then there was *Supernova*, which seemed to be one of the few sources of true contention within the group. The piece, authored collectively by Ale, Juan José, Nahum, and Tania, was meant to replicate the explosion of a star using a traditional Mexican piñata, covered in shiny silver foil with white streamers dangling from its seven points. After smashing a piñata in zero gravity, its candy would "fly to all possible directions, behaving as a supernova." The piece was meant to be a "metaphor of cosmic traditions." The cosmonauts, Miguel told me, thought it was meant to be a representation of Sputnik.[3] Indeed, the cosmonauts had to do the work of smashing the piñata, because it would have been

FIGURE 11 *Supernova*. Courtesy of Nahum, Space Affairs.

dangerous for one of the artists to do so. The Russians loved the Mexican team but were less than thrilled with the time it took to clean piñata debris out of the plane's nooks and crannies.

However, the mess was not the problem. A few weeks before the flight, on the night of September 26, 2014, in Iguala, Guerrero, forty-three students from the Raúl Burgos teaching college in Ayotzinapa were violently "disappeared" on their way to participate in the commemoration of the 1968 student movement in Mexico City. It was eventually determined that the attack was carried out in the context of a horrifying act of collusion between criminal organizations and state police. The aftermath of the "case of the forty-three," which still has not been completely resolved, rocked Mexico. Then, during the ten days that the artists were in Russia, the first reports describing the grisly fate of the students were published, and the government's role in the violence and its cover-up came to light.[4] Not all members of the group agreed that a celebratory piñata was appropriate during a time of national mourning and political protest.

But for Juan José, it was important to "change the conversation," arguing that "good news is more important than bad news. People don't understand that, that you construct your own reality."[5]

IN THE NAVEL OF THE MOON

There is something abnormal and rough in the Moon; a sacred horror grows, that joins with its face that is its true shadow. I leave it high in its epic universe, and almost immaculate for my verse.

—JORGE LUIS BORGES, 1969

The winning image in the open category of the AEM's 2018 space art contest was a black-and-white ink drawing by Juan Carlos Cuevas Méndez called *Dreams Are Built on Roots*. It features a young girl in Indigenous dress climbing a ladder anchored in a mountainous landscape, reaching up to an anthropomorphous crescent Moon with a rabbit perched in its curve. Clouds and shooting stars provide a backdrop for the Moon, while inside the mountain below, Pakal rests in the roots of the axis mundi or in his spaceship; either way, he awaits a cosmic journey. A ribbon twists around the Moon and the top of the ladder: "In the navel of the Moon, we have always been here." As the artist reminds his public, the Nahuatl word "Mexico" is often translated as "In the Navel of the Moon," combining "Meztli" (Moon) and the locative "-co." Given that the contest was ostensibly promoted as a way for Mexican citizens to share in NASA's Artemis mission to *return* humans (not "mankind" anymore) to the lunar surface, what does it mean to say that "we" (not NASA) have "always been here"? The Moon is Mexico, Mexico is the Moon, suggests Cuevas. It is not "out there," but "here."[1]

Cuevas's drawing points to the ways in which the Moon operates as a unique nexus where different milieux intersect and different forms of worldmaking (and unmaking) occur. In this chapter, we move from In-

FIGURE 12 *Los sueños se construyen sobre las raíces / Dreams Are Built on Roots.* Courtesy of Mario Arreola-Santander, Outreach Director of the Mexican Space Agency.

digenous relations with the Moon to the ways in which the Apollo lunar missions were perceived in diverse ways by Mexican publics, from Mexico's role in cosmic diplomacy in the mid-twentieth century to its adherence to the Artemis Accords in the twenty-first century, from technoscientific lunar missions to artistic engagements with lunar landscapes, and from a neo-shamanic eclipse expedition to a shamanic narrative about astronauts. The Moon becomes a complex ontological subject (a physical satellite, a scientific object, a cosmic being that interacts with humans, a resource for future development, a symbol of technological achievement or failure, a site of diplomatic negotiation, a body, a world) that participates in multiple Mexican milieux that negotiate access to the Moon in diverse, sometimes contradictory, ways.

Different milieux produce distinct but overlapping lunar worlds through their practices, imaginaries, and forms of engagement. The Moon may be a scientific object to be studied, a cultural being requiring relationship, a diplomatic space to be negotiated, a resource to be developed, or a ritual agent in cosmic cycles. These worlds intersect and sometimes conflict, particularly around questions of access, authority, and legitimate forms of lunar engagement. The power dynamics between milieux shape which lunar worlds become dominant, as the scientific/commercial Moon of space agencies and private companies often overshadows Indigenous and artistic lunar worlds. Yet alternative lunar worlds persist and even flourish in the interstices, suggesting possibilities for what Isabelle Stengers calls "cosmopolitical" negotiations between different ways of knowing and engaging with the Moon. This chapter traces how these different lunar worlds are produced, maintained, and negotiated across Mexican space milieux, revealing the Moon as a site where competing visions of modernity, nationality, and human-cosmic relations play out.

One Giant Step . . . for Whom?

ROSA INÉS: *What was your first memory of space?*

JOSHUA: I think it must have been the song about the Moon, the one about the rabbit. Every full moon my mother sang it to me. That must have been it.[2]

ROSA INÉS: *What was your first memory of space?*

ITALIA: Umm, the Moon.

ROSA INÉS: *The Moon? And how did it look?*

ITALIA: Umm, like a big round ball.

MARIEL: *And what do you think it's like out there in space?*

ITALIA: Well, I think it's really big.

ROSA INÉS: *If you could go to the Moon or space, right now, would you go?*

ITALIA: Yes, but only if I could go with my mom.[3]

In 1969, official diplomatic discourses focused on the first Apollo Moon landing as an opportunity for international cooperation. Mexican journalists traveled to Houston as part of the group of eight hundred foreign correspondents who witnessed the event, which also saw the first images of space transmitted live, simultaneously to countries around the world (González de Bustamante 2012, 135–39). According to the memorandum the United States Information Agency (USIA) sent to President Richard Nixon about foreign perceptions of the event, of the six hundred fifty million worldwide television audience, seventy-five million people watched the Moon landing in Latin America, through broadcasting agreements by NASA and COMSAT.[4] The event was also relayed to radio stations, thus reaching a much wider rural audience. The memorandum notes that:

> some papers in both the industrialized and underdeveloped world . . . mentioned the contrast between the billions spent for space exploration and lack of success in dealing with urgent problems facing humanity here on Earth—but many . . . answered such criticism by noting that this great undertaking of man does not hinder human progress, but in the long run helps humanity marshal its talents and resources in solving age-old problems.[5]

However, the USIA was happy to report to Nixon that "evidence of the profound impact of the moon landing from all corners of the world is copious and often moving." Among other expressions of global enthusiasm, the USIA notes that "school children in Bavaria and students in Mexico were excused from classes that day."

President Gustavo Díaz Ordaz, one of the seventy-three world leaders who had contributed messages etched on a small silicon "goodwill" disk to be left on the Moon's surface, watched the event on a color television in Los Pinos, although most of the 4.5 million television sets in Mexico, half of them in Mexico City, broadcast the event in black-and-white. Those who did not have a television, or who were away from home at the time, watched the coverage of the Apollo 11 mission in shopping centers, hospitals, and the international airport.[6] Miguel Alemán Velasco, son of the former President Miguel Alemán Valdés, ex-governor of Veracruz, and the director of Telesistema Mexicano's news division, covered the

event from Cape Kennedy along with thirty-eight other Mexican journalists and government representatives.

> This is a historic morning. The flight to the Moon has been compared, for
> its importance, to many things, from the voyage of Christopher Columbus
> to the flight of the Wright brothers. . . . The historians of the future will
> surely say that, on this day, a new phase for humanity and the universe
> began.[7]

Jacobo Zabludovsky, covering the Moon landing for Telesistema (now
Televisa), told his audience to "stop your watch and make sure to preserve
this moment in your memory! This has been the instant, the fraction of
a second, the lightning bolt that divides two epochs like the split in an
abysm." Channel 8, an independent television station that later merged
with Televisa, also broadcast the event, with famed Mexican comedian
Cantinflas narrating, as part of a series of programs about the Moon
titled "Around the Moon in 8 Days," a reference to the 1956 film version
of *Around the World in 80 Days*, in which he had starred. A few years
later, Cantinflas would tell the story of the Moon landing again on his
animated *The Cantinflas Show*, referring to the "Conquest of the Moon"
as the greatest human adventure of all time. Just after the cartoon Neil
Armstrong plants the U.S. flag on lunar soil, the cartoon Cantinflas plants
a small Mexican flag on a lunar crater saying, "Let all men of good will
come to the Moon." His character then visits the U.S. Congress to ask
the government to abolish "The Law of Gravity," which they happily do,
allowing Cantinflas to float back up to the Moon. "*Jóvenes, ¡querer es
poder!*" (Young people, where there's a will, there's a way!)[8]

Individual Mexicans also responded to the first Moon landing in very
different ways, incorporating it into local frameworks of reference. As
might be expected, those of my interlocutors at the AEM who are old
enough to have witnessed the event remember watching the Moon landings on television with great nostalgia, and some cite it as the start of
a life-long interest in outer space. In a biography of her then-husband
Guillermo Haro, founder of the INAOE, an expert in flare stars, and
one of the discoverers of Herbig–Haro objects, Elena Poniatowska (2014,
301) describes watching the Moon landing with Haro and their children,
writing that, "at the most critical moment of the Cold War and when the

conflict in Vietnam showed the world the worst face of the United States, the words of Neil Armstrong move the entire world."

As in other countries, the Mexican advertising sector also took advantage of the Moon landings. As well as companies promoting mattresses, cars, soft drinks, and watches, Cemento Tolteca, one of Mexico's largest cement producers, appropriated the event to promote their product. One of their publicity images, showing the hand of an astronaut holding a (concrete?) Moon in his hand, linked spatial and terrestrial technologies:

> Man has planted his foot on lunar soil. It is the conquest of centuries, the splendid effort of genius and human greatness of all time, that finds its full expression in the impetuous development of modern science and technology. The most advanced science and technology have also produced concrete, the construction material of the contemporary epoch. The main ingredient of concrete is cement.[9]

The Apollo astronauts visited Mexico City in September 1969, the first stop on their goodwill world tour of twenty-four countries. Pictures show throngs of people surrounding the car that carried the "conquerors of the Moon" (who were wearing charro hats) through the streets of the capital. The newspaper *El Universal* reported that:

> Women wearing the colors of the Mexican flag gave them flowers, and they got into a black Lincoln that then-President Gustavo Díaz Ordaz placed at their disposal. They drove off toward the Historical Center and the hysteria took the form of a traffic jam, the trip became almost as difficult as the space journey itself. . . . As they passed through Tepito, sellers gave them gifts, as did dancers and women in regional costumes, the space men responded 'gracias' in clear *agringado* Spanish.[10] (Reyes 2017)

In an article commemorating the fiftieth anniversary of the first Moon landing, reporters for *El Universal* interviewed Mexican citizens in the street, asking how they felt about the occasion. While some were enthusiastic, not everyone was convinced of the event's transcendence, or even of its authenticity. One sixty-year-old woman from Azcapotzalco commented that when she was in the fourth grade in 1969 she was supposed to watch the transmission for a school project, but lacking a television or radio, she couldn't do her homework. A newspaper seller from down-town Mexico City watched the transmission in black-and-white but felt that "it wasn't a great leap for humanity, although it was a good research project, because the Russians had already advanced a great deal in space" (Castro Sánchez 2019).

Cold War Cosmopolitics

The 1960s in Mexico were characterized by contradictory forces and events. On the one hand, the decade was marked by violent governmental suppression of social movements like the Tlatelolco massacre. At the same time, the nation was rapidly becoming more urban and "modern," as this period saw the construction of housing developments, transportation, and communications infrastructure, as well as the satellite transmission of global events like *Our World*, the Olympic Games in Mexico City, and, of course, the Moon landing. The Mexican government presented itself on the international stage as a regional leader and diplomatic intermediary, whose goals centered on world peace in the context of the global Cold War. The Treaty of Tlatelolco banning nuclear weapons in Latin America and the Caribbean was signed in Mexico City in 1967, the same year as the Outer Space Treaty, an international agreement meant to place legal limits on human activity in space. Like the Treaty of Tlatelolco, the Outer Space Treaty (formally the Treaty on Principles Governing the Activities of States in the Exploration and Use of Outer Space, Including the Moon and Other Celestial Bodies) responded to fears of a nuclear arms race, prohibiting the placement of any weapons of mass destruction "in orbit or on celestial bodies" or "in outer space in any other manner." The Outer Space Treaty, whose fundamental principle is the peaceful use of the Moon and other celestial bodies, also requires states to assume liability for damage caused by their space objects, avoid harmful contamination of space and celestial bodies, and claim responsibility for national space activities carried out by governmental and nongovernmental entities. According to the treaty, all states should have equal access to outer space, no nation can claim sovereignty off-Earth, and "the exploration and use of outer space shall be carried out for the benefit and in the interests of all countries and shall be the province of all mankind."[11] The Mexican government ratified the treaty in 1968, cosmically scaling its regional diplomatic role.[12]

Mexico was one of the eighteen founding member nations of the Committee on the Peaceful Uses of Outer Space, created by the United Nations in 1957, just after the launch of Sputnik. In addition to the Outer Space Treaty, Mexico is a signatory of the other four agreements implemented by the UN, including the controversial "Moon Treaty" (The Agreement Governing the Activities of States on the Moon and Other Celestial Bodies) of 1979, which, so far, no spacefaring countries have ratified. The Moon Treaty goes further than the Outer Space Treaty in declaring the Moon "the National Heritage of Mankind," which, in principle, would limit human activities that could damage the satellite, privilege scientific research, provide a regime for international cooperation (unspecified), protect the Moon's environment, and promote the "responsible exploitation of natural resources" and "the equitable sharing by all state parties in the benefits derived from those resources."[13] The Moon Treaty, while maintaining that the Moon should be understood in human terms, also insists on the principle of global equality with respect to access and benefits, and Mexico's adherence to the agreement is consistent with the nation's stated interest in avoiding international conflict and promoting international cooperation.

In this context, Mexico was a consistent participant in the performances of international Cold War cultural diplomacy known as world fairs. The artworks and museum curations created for these exhibitions were meant "to unite nostalgia for the past with enthusiasm for the space race, which coexisted with the utilization of culture and the arts as a tool of the so-called 'soft power,' visually constructing a series of symbols and democratic principles" (Cruz, Garay, and Velázquez 2021, 14). Mexican cultural promoters—policymakers, artists, and architects who had been instrumental in designing and building the edifices of modernist Mexico within the nation's borders—were interested in world fairs as opportunities for worldbuilding, chances to present Mexico internationally as a modern, cosmopolitan country, with a glorious civilizational heritage.

Organizers of the 1968 HemisFair world exhibition held in San Antonio, Texas (called "the point of entry into Latin America" in promotional materials), programmed their event to accommodate that year's Olympic Games to be held in Mexico City (Cruz 2021, 79), which itself would include a "Cultural Olympics" incorporating many elements of midcentury world fairs: spectacles of ancient and modern art and culture and expositions about advances in science and technology, particularly nuclear power and space exploration. For HemisFair, U.S. organizers commissioned a mural from Juan O'Gorman to be titled *Confluence of Civilizations*, the theme of the fair, which provided an opportunity for O'Gorman to expand on the themes of technical optimism and cultural heritage he had explored in his murals for the UNAM's national library. Like many works characteristic of the Mexican School, *Confluence* portrayed oppositional forces that connect in the center of the mural in an aspirational new synthesis. In this case, the left side of O'Gorman's mosaic represented Mexican history, emphasizing pre-Hispanic culture, colonial *mestizaje*, and Indigenous contributions to agricultural science, underneath the Nahua glyph of the rabbit in the Moon. In parallel, the right side of the mural showed the transformation of Western science since the Greeks (imagined as the origin of "the West"), culminating in an astronaut underneath a sky inhabited by spaceships and satellites, pierced by a rocket rising out of the nucleus of an atom. The gendered division between "culture" and "science" is manifest in the figures of a Mexican woman and an Anglo man (an Eve and an Adam) holding hands over the image of a child, presumably the union of the two traditions, that connects the two sides of the mural. This visual content echoed the Texan organizers' marketing slogans that claimed that HemisFair "would combine the joy and spontaneity of the North American West with the adventure and magic of the space age" (77).

Mexican organizers, however, were more interested in moving beyond the social realism of the Mexican School represented by O'Gorman. Having already participated in the 1967 Montreal Expo, Rufino Tamayo was again invited to contribute a more abstract (and therefore more "modern") mural to the official pavilion at HemisFair. Despite its more "modern" style, Tamayo's *Brotherhood*, a thirty-five-meter transportable painting depicting a circle of men of different races with their arms interlaced surrounding a bright bonfire in the shape of a flower, was in many ways a return to Vasconcelos's "cosmic race." In 1971, the Mexican government gifted the painting to the United Nations "as a symbol of its entry into modernity . . . despite the country's political reality before and after October 2" (Cruz 2021, 88).

Tamayo was the most consistently "cosmic" of the Mexican midcentury artists; his paintings depicted the "Mexican Man" as a universal figure, evoking a "nationalism that was more poetic than political" (Torres 2021, 74). Works such as *Gazing at Infinity* (1932), *Total Eclipse* (1946), *Dog Barking at the Moon* (1942), *Women Getting to the Moon* (1946), and *Man Facing Infinity* (1950) were evidence of his preoccupation with the relationship between humanity and the vast universe. By the 1950s, Tamayo's growing interest in astronomy and astronautics had begun to inform his paintings. Rather than an abstract circle, the Moon became a landscape pockmarked by craters in Tamayo's post-Apollo paintings. In works such as *Man in Space* (1970), *Men in Space* (1970), and *The Astronauts* (1971), as well as *The Conquest of Space* (1983), a sculpture commissioned for the San Francisco International Airport, Tamayo reflected on the tensions between his passion for space exploration and what he saw as the possible dehumanizing effects of technology (Ávila Jiménez 2010, 54). As the art critic Emily Genauer wrote in 1975, the Mexican artist

> ... sees technology ... in big, mythic terms, as an implacable, devastating force looming over men almost, one gathers, as he feels the gods loomed over ancient Mexico, where they were the source of light, energy, rain, fire—all the gifts men required to live—but also regularly demanded cruel blood sacrifices. Technology to him is the root of present-day man's great dilemma, the cause of his frailty and fear, but at the same time the instrument by which he can rise above himself and his present needs. (24)

Tamayo had been part of a group of six international artists (including Marcel Duchamp and the Brazilian surrealist Roberto Matta) invited to visit the headquarters of NASA "not in order to register his observations in paintings or drawings, but to discuss with them the relationship between science and art on a theoretical level" (Genauer 1975, 27). Years later, after a car accident left him unable to eat solid food, Tamayo's friends from NASA sent him cans of liquid food used by astronauts (Ávila Jiménez 2010, 51). In an interview published few years before Tamayo's death, a journalist asked him about his reaction to the new discoveries emerging from space science.

TAMAYO: I think they are amazing, and also horrific. You can get really scared, thinking that there are astros that are more important than the Earth. The Earth is such a small thing . . . that there must be more life out there on bigger planets. It isn't possible that [there is only life] on such an insignificant thing in the middle of the immense universe.
JOURNALIST: Would you like to take a trip to the Moon?
TAMAYO: Of course I would.[14] (55)

David Alfaro Siqueiros was the only one of the "Big Three" muralists still living when Neil Armstrong imprinted his boot on the Moon. His easel painting *Stratospheric Antennas* (1949) had already hinted at the emergence of telecommunications technology in the national consciousness, while *Space Nostalgia* (1969) presented a "cosmic landscape, ethereal, lunar" (Juárez 2010, 30) and *Earth Seen from the Stratosphere* (1971) foreshadowed the post-Apollo notion of the overview effect by envisioning Earth as seen from a spacecraft. But in his final major work of public mural art, he would project the Moon landings as the culmination of a dream of achieving world peace through technoscience. Painted between 1964 and 1971, the work was originally called *The March of Humanity in Latin America*, although its title morphed into *The March of Humanity from the Earth to the Cosmos*, and subtitled *Misery and Science* (Folgarait 1987, 51). Although he had been invited to participate in several world fairs, the artist declined, arguing that he was too busy with monumental work.

Said to be the world's largest mural, the eight-thousand-square-meter sculpture-painting adorns both the interior and exterior walls of Mexico City's Siqueiros Polyforum. It leads the spectator on a dizzying journey throughout the history of the human struggle, from the failed "Bourgeois Revolution" to the hopeful, yet-to-be-achieved "Revolution of the Future." Images of human struggle on Earth, the "misery" of the mural's subtitle, include depictions of racist violence, worker oppression, militarism, and apocalyptic landscapes. In contraposition, cosmic "science" is evoked through representations of technological achievement and nuclear power, with the possibility of a triumphant, rational, collective, Latin American future represented by, among other elements, the figure of the last Aztec emperor Cuauhtémoc and a rocket launching into space, accompanied by a Soviet red star. Giant male and female figures reach out from each side of the cupola. "The man offers 'science and technology, bringing industrialization', the woman, 'culture', two components that will bring about 'a humanized society'" (Folgarait 1987, 73).

"While . . . in Central America Our Brothers Die"

Coverage of the Moon landing in independent news outlets tended to be more ambivalent than the pro–mainstream U.S. news sources, which often relied on information and images obtained from the U.S. and maintained financial agreements with U.S. media companies (González de Bustamante 2012, 114). An example of more critical journalism can be found in the pages of the independent weekly magazine *Siempre!*, known at the time for its support of the Cuban Revolution and other regional social movements, promotion of Mexican diplomacy in Latin America, and anti-U.S. stance (Cabrera López 2013, 49). Leftist intellectuals such as Fernando Benítez (the supplement's founder), José Emilio Pacheco, Octavio Paz, Carlos Fuentes, Carlos Monsiváis, Rosario Castellanos, and Elena Poniatowska often published opinion pieces in *La cultura en México*, the magazine's cultural supplement. In August 1969, several pages of the supplement's interior were dedicated to an explanation of the Moon landings, illustrated with NASA's photographs. But the cover of the supplement was divided into two images: the upper third was dedicated to "The Last Days of Trotsky" (who had been assassinated in Mexico in 1940), while the bottom section had a black-and-white photo of an eclipse and the heading "The Moon and Nine Writers." Inside, the title read "Homage and Profanations: The Conquest of the Moon As it Might Have Been Viewed by Eight Mexican and One Argentinean Writer." In his submission, Carlos Fuentes (1969, 8) alluded to the déjà vu–inspiring trope of conquest:

> At the peak of his power, Nixon-Moctezuma sees in the sky the end of his empire. How to conquer the Moon without violating it? While in Vietnam is being written the modern Iliad, a reverse Odyssey is being accomplished in space, a return of Man to where Man has never been. Armstrong-Jason returns with the fleece—the dead rocks of the Moon—projecting ambigu-

ity and the diachronic continuation of myth.[15] Quetzalcóatl goes ahead to announce the arrival of other men like him. Who will be the Bernal Díazes and Inca Garcilasos of this new conquest?

In his poem "Paschimottanasa," Octavio Paz made a play on words with an asana or yoga pose and NASA, perhaps inspired by his time as ambassador to India, a post he had renounced the year before in the wake of the Tlatelolco massacre. His critique was more direct, along the lines of Gil Scott-Heron's (1970) oft-cited "Whitey on the Moon." Paz (1969, 9) wrote: "A dark sky / is not a dark garden / Radiation / The appearance of Man / in the stony desert / While they wake up on the Moon / in Central America / our brothers die." Carlos Monsiváis (1969, 10) began his text:

> FADE IN. Me and the Moon in 1969. Without a critical attitude, no conquest of the Moon is possible. What else can I say? Nothing, except that the trite does not recognize stratospheres, and the same happens in the Janitzio of Agustín Lara and the Ballet Folklórico as in the transmissions, obviously influenced by comic books, from Apollo 11. And do not think that in these notes we speak with the underdeveloped resentment of those who were offended because when the lunar module descended Neil Armstrong did not shout "¡Viva México!" or kiss an image of the Virgin of Guadalupe.

Monsiváis ended his reflection on the colonization of the Moon with a postcolonial affirmation: "If you confront me with your Neil Armstrong, I will respond with my Franz Fanon" (10).

So Far from God, So Close to NASA

In 2023, NASA administrator Bill Nelson and his staff visited Mexico as the first stop on a diplomatic tour of Latin America that many of my interlocutors at the AEM interpreted as a reminder of the importance of U.S.–Latin American spatial cooperation in the face of an increasing Chinese presence in the sector, and indeed China and Mexico have established "integral strategic associations" (Frenkel and Blinder 2020, 3).[16] As one AEM representative told me in confidence, "It feels like the U.S. only pays attention to us when they think China is getting involved."

In a reception held for NASA officials at the home of the U.S. ambassador to Mexico during this visit, the Mexican Secretary of Agriculture proclaimed that "Mexico and the United States were no longer 'Good Neighbors,' but 'compañeros *de cuarto*,' roommates."[17] Cue smiles, nods, and some uncomfortable laughter.[18]

In one example of this ambivalence, uncertainty surrounds the possibility of developing launch sites for satellites within Mexico, one of the AEM's long-term projects. I was told that several sites had been discussed, and several countries were interested, but that sites near the border were problematic, as the U.S. government had expressed concerns over national security. A different interlocutor, who works in the development of the commercial uses of space at the AEM, told me that during a visit to the U.S. to "test the waters" for the construction of a launch site in Baja California he was told that "of course, you are a sovereign country and can do what you want," followed by an implied "but . . ."

Artemis

As I mentioned at the beginning of this chapter, the "Mexico to the Moon" space art contest sponsored by the AEM was held in the context of the Artemis Accords, nonbinding, bilateral agreements between the U.S. and other world governments promoting collaboration on outer-space activities. In 2021, during a press event that included the Mexican ambassador to the U.S. and the U.S. ambassador to Mexico, the governor of the state of Hidalgo, members of the Mexican Senate, the Secretary of Foreign Relations, and Chicano astronaut José Hernández, Mexico became the fourteenth signatory of the Artemis Accords. Some have questioned whether the Moon Treaty and the Artemis Accords are congruent, given that the Accords, while still emphasizing the peaceful uses of outer space, are focused on the right to extract and use celestial bodies' resources for commercial purposes (Tronchetti and Liu 2021, 244). And, while the Accords are formally bilateral, they remain heavily weighted toward U.S. interests. But for many of my interlocutors, by signing the Accords, Mexico has re-entered the space race. As the Secretary of Foreign Relations declared fifty years earlier, "We were spectators, now we will be participants."[19]

Announcing the signing in the AEM's online publication, Carlos Duarte (2021) exulted, "We are joining an exclusive club made up of the United States' traditional partners in the field of outer space, such as Canada, Japan, the United Kingdom, and Italy, as well as emerging countries like Brazil, Poland, South Korea, and the United Arab Emirates." Duarte celebrated the impact that the adhesion to the agreements will have on the Mexican space industry and the potential for generating "an avalanche of opportunities, new projects, and new companies that could change us forever." At the same time, he acknowledged that the apparent conflicts between the Artemis Accords and the Moon Treaty would have to be resolved eventually.

Another issue to be resolved, wrote Duarte, is the lack of legislation around the use of outer space. Indeed, space law, which had been one of the areas developed in the era of the CONEE, is again on the rise in Mexico. Much of the legal discussion revolves around international law having to do with the regulation of satellites and their frequencies (Álvarez 2023), but the topic of commercial regulation has become increasingly pertinent. In 2023, the federal House of Representatives (*Cámara de Diputados*) approved a measure that would reform the Mexican constitution to establish outer space as "a priority for the nation" and give the legislature the power to formulate laws regarding "the colonization of outer space" (López Velarde 2023, 30–31). According to my interlocutors, the reform is intended to make the commercialization of outer space less complicated in legal terms. However, although the measure has passed one legislative chamber, it has yet to be approved by the Senate. Apparently, Hacienda (the Mexican IRS) is not convinced by the proposal. "They think they'll have to spend money," one representative of the AEM told me, "But they don't understand what the impact will be."

Beyond its national borders, Mexico continues to participate actively in COPUOS (at least as of 2024). Lawyer Rosa María Ramírez de Arellano y Haro, who became the first woman and the first Mexican to chair the committee during its 2018–2019 session, was also the coordinator of the AEM's division of International Affairs and Security in Space Matters until the AEM's merger with MEXSAT in 2025. But cosmic diplomacy is complicated for this country, caught as it is in the middle of frictive national and geopolitical interests. As we have seen, the pull of the U.S.

is particularly strong, forcing contemporary Mexican space institutions
to try to navigate between competing lunar worlds.

Colmena: "The Mexican Conquest of Outer Space"

ROSA INÉS: Do you think Mexico can or should participate in space exploration?

CARLOS: Well, we've only had contact since Neri Vela went to space, and
well, that's over, more than thirty years ago, 1985, but I still feel it
would be worth it to be closer to that environment because, ultimately,
it would be a solution to the things that we are missing here. They're
researching materials on Mars, but it's really far. So is the Moon, but
we could be surprised.[20]

Fifty years after Jorge Luis Borges wrote of leaving the Moon in the sky,
"almost immaculate for my verse," I talked to another Argentinian who,
by contrast, *wanted* to leave his mark on the lunar surface. When I visited his laboratory in 2019, Gustavo Medina Tanco, whom we met in
chapter 2 as the head of the UNAM's space instrumentation laboratory
and developer of satellites, was working on a more spectacular project
meant to catapult Mexico into the elite international group of lunar explorers. Some of the LINX team members were working on projects
such as the NanoConnect-2 satellite that would eventually be launched
from India, or the creation of LANAE, a center dedicated to the launch
of stratospheric balloons in Hidalgo. But others dedicated their time to
Project Colmena, or "beehive," an assemblage of nine small robots (five
would eventually launch) that had already booked passage on a flight to
the Moon with the U.S. company Astrobotic. However, owing to a series
of delays common in the space industry, the mission would not launch
until 2024. "What can we do?" Medina said to me, rhetorically. "We can't
put any pressure on them because we're just a small client. Maybe if we
were NASA. If you pay, you can get to the Moon quickly. Our trip is more
like 'traveling on a donkey.' It will take months."

The robots I saw in 2019 looked to me like pieces of a Lego construction set. Each identical robot is a disk with wheels set into its center
and a ridged circumference, a little under five inches in diameter and

weighing a little over two ounces. On that first visit, Colmena's robots were called *tepoztli*, "metal things," but the name was eventually dropped as it would be difficult for English speakers to pronounce. (The mission itself had been called Meztli, "Moon" in Nahuatl, but that name had already been abandoned.) Next to the robots was a model of Astrobotic's Peregrine module that would house them: a golden half-cylinder with some support tubes on an octagonal base. A sign propped up next to it said, in large letters, "*NO TOCAR.*" Next to the module was one of the looms that were being used to weave the robots' protective covering, laced with a "very expensive" thread able to resist radiation, UV exposure, and temperatures from −150 °C to +200 °C, as well as to support pressures of up to 3.2 GPa. On Facebook earlier that year, Medina Tanco had attempted to bridge scientific and cultural lunar worlds by posting images of his students using the looms, along with an image of the famous Zapotec weavers of Teotitlán del Valle in Oaxaca: "The fact that five hundred years after the conquest, a small piece of Mexico's cultural traditions will find a resting place on the Moon's surface, along with twenty-first-century technology!"[21] I also got to see the robots' analog playground, which looked like an ordinary sandbox but was filled with ground basalt rocks meant to mimic lunar regolith, explained by Oliver Morton (2019, 136) as the result of "billions of years of bombardment" from which Earth is protected by its atmosphere, and by the Moon itself. As we left the laboratory, Medina Tanco informed his students that TV Azteca would arrive the next day to report on Colmena, asking them to organize the lab so that it would look good on camera. "More gadgets everywhere," he directed.

Colmena's goal was to show the possibility of building structures on planetary surfaces using swarms of self-organizing miniature robots "inspired by nature." According to Medina Tanco's vision, thirty years into the future, Mexico would have developed microrobotic technology for a variety of applications, including the obtention of minerals and water from lunar dust or rare earth mining on asteroids that, according to proponents of outer-space extractivism, could potentially have a market value of trillions of dollars (Jamasmie 2016). Morton (2019, 200) describes this view of outer space as providing endless resources, "endless spacey cake and endless Earthly eating," as a "High Frontier that would never close. It would just get higher and higher. . . . Cosmism as capitalist self-improvement. More and more Moonshots: even fewer have-nots."

The robots were designed to do two things: first, to organize themselves on the surface of the Moon, linking their individual solar panels into a larger and more potent solar panel, and second, to measure the effects of lunar dust on the robots' operation. Underlying these techno-scientific objectives was the more mediatic goal of convincing both the national citizenry and the international space community that Mexico has the potential to become a space actor. The use of Indigenous names and cultural references was part of the mission's national marketing strategy to connect Mexican identity with cutting-edge technology. There was originally talk of holding a poetry or art contest so that Mexican students might send their creations to the Moon, although this eventually proved to be too complicated.

The discourse of a "Mexican conquest of outer space," repeated unironically in national news outlets, promised to redress the national trauma of a historic conquest.[22] In the media blitz before and during the mission, the public read enthusiastic reports like the following from the magazine *Wired*:

FIGURE 13 Colmena. Photo by the author.

"Mini robots dug up on the Moon" will be the headline of a report about an astroarchaeologist who will discover the remains of the Space 4.0 Anthropocene, the era we are living in now. Upon investigating, they will discover that the technological bugs are Mexican and that they were part of the first Terran bid to take micro-robotics to interplanetary space. (Castro and Miguel 2024)

How to Do Things with Failure

In 2022, I went back to the LINX to help translate during the mission's first simulation, which would require students working at the lab to communicate with the Astrobotic team in Pittsburgh as they rehearsed the four separate sequences that would make up the mission's operation. "Everything sounds like you're in a tin can. And besides," Medina laughed, "We need you because they speak Texan." (They didn't actually speak "Texan," but "Pennsylvanian.") The lab had been provided with a PowerPoint document by the company showing each step and the proper responses. For me, programming languages are completely foreign, and the student chosen to communicate with Astrobotic spoke English very well, so my input was minimal, but it was fascinating to get some insight into the process. First, the students practiced the self-test and then the deployment sequence. We waited, imagining what the robots were doing on the Moon. Then they sent the command for the "clustering" sequence. A minor error in the code was quickly corrected. Finally, they sent the command for the science test, and the simulated mission was completed and successful; the tepoztli were moving around on the lunar surface and gathering data.

Unfortunately, the mission's "real" success would be more limited. The robots never reached the Moon, although they did "travel further from Earth than any other technology made in Mexico." After several years of hearing "this month, this year, soon . . ." Astrobotic's Peregrine One mission launched from Cape Canaveral on a United Launch Alliance rocket on January 8, 2024, carrying Colmena, another rover designed at Carnegie Mellon University, and two smaller rovers developed by Japanese and Chilean companies. Other payloads included scientific instruments developed by NASA (the first payloads sent by the agency under its Commercial Lunar Payload Services, or CLPS, program) and German researchers, as well as a series of time capsules provided by universities and companies from various countries.

The launch itself was successful, and the separation of the solid rocket boosters, followed by the separation of the first stage, the upper stage burn, and the separation of the Peregrine lander all occurred without incident. The plan was for Peregrine to take a trajectory that would allow it to enter lunar orbit in forty-six days, with a Moon landing projected on February 23. However, a problem with its propulsion system prevented the mission from leaving an elliptical Earth orbit. Rather than attempting a crash landing on the Moon, Astrobotic decided to return the lander to Earth after traveling around 234,000 miles away during its almost seven days in space, allowing it to burn up in the atmosphere to avoid creating space debris. "By responsibly ending Peregrine's mission, we are doing our part to preserve the future of cislunar space for all."[23]

Medina Tanco insisted that Colmena was not a failure; rather, the fact that his team was able to establish contact with the robots and monitor their operations while in orbit meant that the mission was a technological and scientific success. "They were functional in even more extreme temperatures than we had originally planned, although we couldn't study the effects of lunar dust. . . . And the miniature catapult we had designed to launch the robots onto the lunar surface also worked, showing our capacity for innovation."[24] LINX sent technology developed in Mexico "400,000 kilometers into space, into lunar orbit, joining a very select club." He admitted that seeing the lander burn up with his robots inside was tough on him and his students, but that he had "chosen to focus on the voyage, not the death of the module."

Medina Tanco was concerned that the media and the public would interpret Colmena's failure as yet another instance of the typical "of course, the national football team lost, again."[25] Therefore, he attempted to "spin the narrative" (his words) and frame the mission in terms of normal, and even positive, technological innovation failure, a *falla* rather than a *fracaso*. Echoing tech culture's discourse that innovation can only happen if innovators are willing to fail, Medina Tanco insisted that failure is a part of "doing outer space," thus spinning "failure" into a vital part of being a space-faring nation, rather than a version of what he described as the self-perpetuating myth of the idea that "we are never able/allowed to succeed."

Convincing the AEM, the UNAM, and the public of Colmena's success was complicated. According to Medina Tanco, the UNAM had wanted to keep the news of the mission's technical problems quiet for a week, but he had insisted on communicating with the media from the beginning: "We can say that at least 75% of our objectives for this mission have been achieved, which has created the future path of Mexico toward the Moon" (Lagos 2024). And "we have shown that Mexico is much more than a *maquiladora*. Mexico can create; Mexico can innovate. . . . Innovation means failing. If you don't fail, you aren't doing anything innovative."[26] The headlines were adjusted from "Colmena: the first Mexican mission to the Moon" to "Colmena: the first Mexican mission to reach lunar orbit" and "Colmena: the first Latin American mission to cislunar space."

On social media (#colmena, #linx, #luna), most reactions were supportive and encouraging: "Congratulations! Constructing the basis for future research!" "I'm crying happy tears! Bravo, Colmena!" "And they said that we Mexicans had no talent for developing technology!" "Mexico is badass!" "Your achievement makes me feel like I'm on the Moon, almost . . ." and "GOYA!" (the UNAM's cheer). However, some responders took the "success" narrative with a grain of salt. "Didn't you run out of fuel?" and "When you build a rocket, let us know." Other posters framed their critique more ideologically: "Wanting to go to the Moon is neoliberal, capitalist, and does nothing for the people."[27] Even some of my interlocutors who believe that failure is necessary for innovation were skeptical of the failure-as-optimism narrative: "It's one thing for NASA or SpaceX to fail; they can start again. But we usually only get one chance." One writer concluded that "the Colmena Mission has been added to the histories of *ya merito* [an "any time now," but probably not soon] in México" (Vallejo 2024).

Carroll et al. write that there is often a discursive slippage between the idea that "things" fail, that objects do not always do as we wish, and failure as a moral phenomenon, with consequences for social relations. They understand failure as "a moment of breakage between the reality of the present and the anticipated future [that] carries moral gravity as what *ought* to have happened, what *should* be the case, has not come to pass" (Carroll et al. 2017, 2). In Latin America, material failure is often tied to the failure of the state; historical experiences of state corruption and neglect have resulted in infrastructural breakdown causing outrage, but rarely surprise.[28] In the case of space technology, the idea that, in contrast to the U.S., Mexican companies do not have the luxury to fail is both a recognition of economic realities and a critique of both the lack of state investment in outer space and of the inequities that characterize global geopolitics.

Outside Mexico, Colmena did not get much media coverage. Peregrine One was most well known for being the first commercial Moon mission and the first CLPS mission for NASA, as well as for its controversial inclusion of human remains, part of payloads contracted by the U.S. companies Elysium and Celestis, which caused the Navajo Nation to launch a formal protest. "It is crucial to emphasize that the moon holds a sacred position in many Indigenous cultures, including ours," argued Buu Nyrgen, the Navajo Nation's president. "We view it as a part of our spiritual heritage, an object of reverence and respect. The act of deposing human remains and other materials, which could be perceived as discards in any other location, on the moon is tantamount to desecration of this sacred space" (Spry 2024). In its response to the formal complaint, NASA recognized the symbolic importance of the Moon on the one hand while, on the other, denying any responsibility for payloads sponsored by private companies. The manager of the CLPS program explained, "We don't have the framework for telling them what they can and can't fly. The approval process doesn't run through NASA for commercial missions" (Tingley 2024). Another NASA representative argued that these early commercial missions would allow the agency to learn about issues that will require regulation in the future. "As time goes by, there are going to be changes to how we view this, or how industry itself maybe sets up standards or guidelines about how they're going to proceed" (Tingley 2024). However, "industry" does not seem to share NASA's or the Navajo Nation's concerns. Ignoring the complexities of cultural belief and practice, the CEO of Celestis decried the potential for the imposition of "religious" influence on outer-space activities. "No one, and no religion, owns the moon, and were the beliefs of the world's multitude of religions considered, it's quite likely that no missions would ever be approved" (Tingley 2024).

Given its history of cosmic diplomacy, as well as its constitutional recognition as a pluricultural nation, Mexico is ideally positioned to contribute proposals for a legal framework around outer-space activities that would include a principle of respect for cultural diversity. As I listened to a panel on the proposed constitutional reform at the Senate in 2024, I talked to Gustavo Cabrera, Mexico's ambassador to the ALCE. I told him I thought that focusing on the commercial uses of outer space limited Mexico's capacity to act in its regional diplomatic capacity, offering alternative ways to think about human activities in outer space, taking into account recent dark-sky legislation (see chapter 6) and historical considerations of the interests and rights of Indigenous populations. Cabrera, an anthropologist by training, agreed, although he thought that the legal vision should include the perspective of popular culture in general, and not just Indigenous populations. At any rate, as we saw above, Mexican space law is currently focused on creating a legal environment that will facilitate commercial activities in space.

I also asked Medina Tanco what he thought about the potential of Mexican legal proposals. He is firmly convinced that the only way to have a voice in what happens in space is to have technology in space. "Picture the expanse of frozen water at the Moon's poles," he told me. "If you don't stick your straw in, you won't have access." And future Colmena missions are already in development: Colmena 2 (originally called Moon-Worm), which will consist of more "organic" robots, without little wheels and motors, and Colmena 3 (originally Moonscouter), which will deploy multiple robots that will help develop technology for Moon and asteroid mining. Asteroids are "an infinite resource," Medina told me. "And the point isn't to develop a gadget but an entire economic area."

An Interplanetary Simulation

The Great Altar Desert in northeastern Sonora, bordering the Sea of Cortés just south of the Arizona border and a few miles away from the tourist and retirement destination of Puerto Peñasco, is known for its biodiversity. The Pinacate Biosphere Reserve, named a UNESCO World Heritage site in 2013 and located within the desert's borders, is home to 805 plant and animal species, dozens of which are endangered. The land is also part of the territory inhabited by the Tohono O'odham, or Papago, who locate the mythical origin of the world in *Shuk toak*, the Pinacate mountains.[29] El Pinacate's volcanic cones, active dunes, and concentration of maars (shallow volcanic craters) give it a decidedly lunar feel, a comparison which, as we saw in chapter 3, did not go unnoticed by NASA. From 1965 to 1970, astronauts from the Apollo program trained for their Moon missions in the biosphere reserve; according to Mario Arreola (2017), the longest training session took place before the Apollo 14 mission, as the region bore marked similarities to the Fra Mauro highlands where the lunar landing would take place. In the wake of the signing of the Artemis Accords, the AEM has shown interest in reviving El Pinacate's lunar credentials and perhaps expanding the project to include training for human missions to Mars. However, community resistance and the presence of organized crime have complicated plans for the future of an extraterrestrial Pinacate.

The artistic and curatorial platform Terremoto ("Earthquake") has also developed plans for El Pinacate. In 2024, members of the collective designed "Interplanetary Simulations" (*Simulaciones interplanetarias*), an artistic residency in the desert meant to "reflect on space exploration, planetary imagination, and the possibilities of the cosmos in the face of

FIGURE 14 Apollo astronauts in El Pinacate, 1970. Courtesy of NASA.

a desolate future on Earth."[30] Five artists, from Chile, Brazil, Argentina, Costa Rica, and Mexico, were selected from over nine hundred applicants for the residency, which was funded by several international organizations, including The Andy Warhol Foundation for the Visual Arts and the Brazilian Embassy in Mexico. The original idea was to begin the experience with a two-week artistic analog mission in El Pinacate, followed by a period of studio/laboratory work in Puerto Peñasco, and, finally, a phase of editorial content creation.

Helena Lugo, curator and director of Terremoto, had been thinking about utopias for a long time before imagining "Interplanetary Simulations." A few weeks after the experiment in El Pinacate, she talked to my friend Marcela Chao of Marsarchive.org (we'll see her in the next chapter), who had accompanied the group to Sonora. In their discussion, Lugo mused on the gap between promise and reality, and, in a dissonant echo of the discourse of failure touched on earlier in the chapter, in which technoscientific innovation requires "fallas," she breached the topic of "utopic failure" as the result of technoscience's broken promises.

> There was this incredible moment during the twentieth century, which has a lot to do with the space race. Humans reached the Moon, you know? And it wasn't just about these super tangible technological advances, but this collective emotion, that technology, capital, science, were going to build the world we always wanted. And I think the twenty-first century is like the total disenchantment or the recognition of all the utopic failures that the twentieth century promised. We still don't have flying cars, or robots like Robotina [Rosie, the robot maid from *The Jetsons*]. I mean, all of the promises of technology, which have been realized in many ways, but which don't live up to the imaginaries that we once created through science fiction, about what life would be like now. That, and a generational disenchantment, generations that live precariously thanks to late capitalism that is devouring everything it promised to build. . . . "Utopia" means "no place." It was Thomas More's joke.[31]

That said, the construction of better worlds is important as a process and a provocation, she continued, and art is what makes them possible.

"Interplanetary Simulations" was born from a collaboration between Helena and Ana Cristina Olvera, a science communicator who works for the AEM. The idea was to create the first artistic analog mission based on the idea of a new world, a blank canvas, that is both inspiring and impossible. They had chosen El Pinacate for its incredible landscape and because they were captivated by the idea of recreating NASA's analog missions of the 1970s, but with a radically different premise: "an exercise of radical imagination." "And the landscape does give you a sense of being on another planet. . . . But it's a project from Latin America, for Latin America, where we wanted to explore our geopolitical and affective realities and to visualize other, global problems, but always from the Global South." The artists participated in dialogues with experts in astronomy and planetary sciences, as well as with local residents who helped them understand the region's cultural and historical elements. Each artist is in the process of producing works that have emerged from their own interactions with the territory: short stories, photography series, an atlas, and other artistic interventions.

The analog aspect of the project proved to be complicated, for the same reasons cited by the AEM—organized crime, violence, and territorial disputes, as well as more mundane problems like interpersonal communication and a lack of water in the bathroom. But the project showed how artists can create spacetimes for the co-existence of different lunar worlds. "Although we would have liked to maintain the fiction of the planetary, the extraterrestrial, the utopic, ultimately, reality always gets in the way. . . . It's that the truth is you're here, you're on Earth. And these are the problems that we have to resolve here before we go anywhere else." Even so, "there were moments when the landscape revealed itself to you as absolutely sublime," like "when you saw moonrise and sunset, Venus at dawn, even Elon Musk's train of satellites." Lugo concluded that "it isn't really about imagining life on another planet, but that moment when you say 'Wow, this is the world.'"

Two-Eyed Moon

If God didn't exist, then we human beings wouldn't exist, and the world would be dark, without light. The great kings or gods of that time helped God in his work, so that the stars would shine in the sky. At that time there were two kings. One was proud and domineering. The other was humbler and had pimples on his face. The domineering one laughed at the humble king. And dear God said, "You are going to be the Moon, you are going to be the Sun." The proud king shouted, "But, why! I am the rich one, I am the good one. How am I going to be the Moon?" God told the humble king, "You, pimply one, you are going to be the Sun, that's why your face is like that. You are poor, but you're going to be the Sun." They had to do what God said. "Obey, do what I will ask you to do. Prepare a great bonfire, very great." Then God told the proud king: "You first. If you want to be the Sun, get in the fire." He began to run, but at the edge of the bonfire, he said, "No, not me." Then the poor king, the pimply one, was the brave one that, without thinking, ran and jumped in the bonfire. The proud king found his courage. "If he, who is poor, jumped in the bonfire, why not me?" He ran and threw himself into the fire. In that instant, an eagle emerged from the bonfire carrying in his beak a golden coin that shone brightly and climbed to the sky; that's why he is the Sun. The other king came out transformed into a *tigre* [jaguar] that bounded up carrying a silvery white coin; he was the Moon. So God saw that the kings obeyed him. That's why even today when it is very sunny, kids get pimples on their faces. Measles and chicken pox are also from heat, because the Sun is the pimply one. (Lorente 2023, 218)

Anthropologist David Lorente recorded this origin story in 2015 in the Nahua town of San Jerónimo Amanalco in the region of Texcoco in Mexico State, which is not far from Teotihuacán, the place where, according to the pre-Hispanic Nahuas, the Sun and Moon were originally created. Today, Nahua storytellers talk of God instead of gods, but the main elements of the pre-Hispanic Nahua myth we saw in the first chapter remain: a poor, sick man sacrifices himself willingly to become the Sun, represented by an eagle, while a rich man hesitates before throwing himself into the fire, and is transformed into the Moon, associated with the jaguar. To distinguish the Sun from the Moon, one of the gods slapped the rich man in the face with a rabbit, dimming his brightness, and leaving the animal's imprint on the lunar surface (López Austin, 1996).[32]

As we saw in the first chapter, according to some pre-Hispanic Nahua cosmographs, the Moon resided in the first of a series of vertical "heavens," inhabited by deities associated with lightning and rain (Tlaloc), wind (Ehécatl), and sexual transgressions (Tlazoltétotl). The nocturnal

Moon was thus associated with the cycle of life and death, fertility, and the circulation of water that was fundamental in sustaining agricultural societies, in contrast to the hot and dry nature of the diurnal Sun and its related forces and beings.

According to French ethnologist Jacques Galinier, Hñahñu (Otomí) communities in the state of Hidalgo associate the Moon with the measurement of time, linking the repetition of the Moon's phases to the recurrence of biological, ecological, and cosmic cycles (Galinier 1990, 235). Hñahñu elders say that the Sun will eventually be destroyed in a sacrificial fusion of cosmic energies, but the Moon will continue to be reborn periodically; "it is the astro of eternity, as it is at the same time life and death . . . it assures the continuity of the history of the world" (239). Galinier writes that, for the Hñahñu inhabitants of San Pedro Tlachichilco, the Sun is "a one-dimensional astro," unlike the Moon (251). The one-eyed Sun is masculine and pure, associated with the heat, the daytime, and power of Christ. In an echo of the myth of the birth of the Fifth Sun at Teotihuacán that we saw in chapter 1, the Hñahñu Sun is also poor, honest, and physically blemished.

The Hñahñu Moon is associated with the celestial mother and the Virgin of Guadalupe, but also with Earth, vegetation, mountains, agave, spiral forms, impurity, sexuality, the cold, water and moisture, weaving, the Devil, witchcraft, dogs, rabbits, spiders, erect penises, menstrual blood, psychological disorders, wealth, the life cycle, and death. Unlike the Sun, the Hñahñu Moon has two eyes, a reference, says Galinier, to the astro's duality and sexual ambiguity (Galinier 1990, 256). During a full Moon, both eyes are open, while the new Moon has both eyes closed. The waning and waxing crescent Moons have one eye open and one eye closed. According to Galinier's interlocutors, "the Moon—like all sensible points of the universe—has no tangible physical existence as an object divisible from other celestial bodies, the Earth, humans, etc. The selenitic astro is a privileged place for the manifestations of the fundamental principles that rule the universe" (257). Human bodies are small-scale replicas of the universe, and human social relations are echoes of cosmic relations. Significantly, the Moon's association with wealth is discursively linked to Indigenous spiritual abundance, in contrast with the perceived spiritual poverty of the mestizo world, in what Galinier calls a "symbolic inversion" of the material wellbeing of the dominant society (295).

Similarly, Lorente (2011, 34) argues that, for his interlocutors in San Jerónimo, the cosmos is characterized by "an asymmetrical dualism" in which the solar forces of heat and light are less relevant than the cold, wet lunar forces that ensure the seasonal corn harvest. The cosmos, which Lorente defines as a "mechanism for the circulation of forces," is regulated through ritual actions that structure a "climatic system" that links the Moon, mountains, rainfall, springs, caves, and the ocean in a world inhabited by humans and nonhuman beings that reside in bodies of water or influence the water cycle. A group of "deified human spirits" called the *ahueques*, or "owners of the water," live below the surface of the earth in a world that is socially and materially similar to the human world, "rich but contingent and perishable," requiring periodic renewal (15). The climate system, or *tiempo* (the Spanish word refers to both "weather" and "time"), depends on a series of interactions between the ahueques and *tesifteros*, ritual specialists, "those who know about tiempo." The ahueques and associated deities (particularly Tlaloc, who is also associated in collective memory with Nezahualcóyotl, the fifteenth-century king of Texcoco who built an extensive hydraulic system in the region) cause lightning and hail, considered types of "meteors," to fall from the sky, damaging the crops planted by humans so that they may consume the plants' aromas, or essences (139). But, in a cycle of destruction and generation, they also send rain to nourish the crops, thus sustaining life on the surface. In a play on words with Vasconcelos's "the cosmic race," Lorente calls the actions of the ahueques "the cosmic raid" ("*la razzia cósmica*"). The ahueques may also steal the spirits of unwary humans, causing illness or death. The job of the tesifteros is to negotiate with the ahueques, ameliorate the damage they cause, cure their victims, predict the weather, and solicit rainfall (108). Tesifteros (called *graniceros* or *tiemperos* in other communities in central Mexico) mediate between the nocturnal and diurnal, underground and celestial world from the middle, *tlaltícpac*, the surface of the earth.[33]

The association between the Moon and the water cycle can be found in many agricultural communities in Mexico, expressed in the idea that rain can be predicted by observing the color of the Moon, that planting should be avoided during the new Moon so that the crops are not vulnerable to plagues or hurricanes, and that the Moon can be angered, which also leads to a bad harvest (González Jácome 2003, 488). For Rarámuri communities of northern Mexico, the Moon serves as a guide for both agricultural and hunting practices. Crops planted during the full Moon will grow tall, but will not give fruit, so planting should be done just before or after the Moon is full, which is a good time to hunt as well, because the light will motivate animals to leave the safety of their hiding places. As is the case for Nahua communities, rituality and more-than-human relationality also characterize Rarámuri relations with the paternal Sun and maternal Moon. The path (*camino*) of the Sun in the diurnal sky and the Moon in the nocturnal sky are models for human behavior, as Rarámuris believe that their mission is to "walk well" (*caminar bien*) during their time on Earth like the astros and the ancestors, a notion that encompasses correct thinking, teaching, and actions, and is materialized in physical movement, such as running and dance (Martínez 2021). "Walking well" is an integral praxis through which Rarámuris not only live correctly as people or local communities but act collectively for the good of humanity, making a habitable world (Fujigaki 2020, 11) and fulfilling their mission to serve as the "pillars" that uphold the cosmos itself (Bonfiglioli 2008, 56).

Human action is also required in the event of threats to the order of the cosmos in the form of eclipses, which may be harbingers of disease and violence. Intervention, in the form of positive ritual practice or negative prohibition, is meant to help the Sun recuperate its energy and life-giving heat and to defeat its enemies. For the Rarámuri, failure could potentially mean the end of the world (Fujigaki 2020, 15). Noisemaking during eclipses is a common activity in Indigenous communities. Echeverría writes of a young Nahua interlocutor from Puebla who told him that during an eclipse the women go to church and pray, but also spank their children to make them cry, and induce dogs to howl and donkeys to bray, thereby strengthening the Sun in his battle against the Evil One. In Guerrero, the author reports, adults attach metal cans to children's clothing so that they clang as they run through the streets, and men in Mayan communities in Quintana Roo shoot their guns in the direction of the Sun to scare off its attacker (Echeverría 2015, 378–79). The Tolupán of Honduras also create sonic disturbances, but, according to Javier Mejuto (2023, 54), they do so to warn Grandfather Sun of the imbalance that the eclipse produces so that he might restore order to the cosmos and assure human survival. Both Hñahñu and Purépecha women bind sharp objects against their stomachs to protect their fetuses from birth defects, and men attach red ribbons to growing plants, which are also adversely affected by the weakening of the Sun during an eclipse (Echeverría 2015, 383; Galinier 1990, 256).

But cosmic events are not the only existential danger: criminal violence, excessive extraction of ecological resources, and anthropogenic climate change also represent grave threats to the world(s) inhabited by humans and nonhumans. Masewal (Nahuas) of the mountains of Puebla view climate change "as a moral crisis" to be countered with the observance of "yearly festivities and ceremonial offerings . . . linked to respectful attitudes of remembrance" (Questa 2019, 40). For Rarámuris, walking what they call the degenerative "other road" (*el camino otro*) is associated with individualistic, "disperse" mestizos (and some Rarámuris who have left their communities behind; Fujigaki 2020, 18). Today, walking the generative, correct path modeled by the Sun and Moon is extremely difficult. However, despite the tensions between diverse desires and expectations, the goal is to connect all beings, human and nonhuman, Rarámuris and non-Rarámuris, "by means of complex networks to administrate, through frictions and conflicts, the regenerative (*semáti*) and degenerative (*cháti*)" (20).

As Szerszynski (2019, 204) writes, Indigenous cosmologies, in which "beings exist in and are defined by meshworks of reciprocity and generosity," are "grounded in forms of social metabolism that resonate with them: that involve passing on the 'accursed share' of excess production in moments of gift and festival, rather than reinvesting it in endless, industrial growth." This insistence on responsible and relational interaction with the multiple beings of the cosmos could provide a useful corrective to extractivist technoscience that seems to be leading humanity to both "epistemological and ontological tragedies" (207). In a felicitous echo of Hñahñu beliefs about the one-eyed, one-dimensional Sun and the two-eyed, multidimensional Moon, Mi'kmaw elder Albert D. Marshall and other Indigenous scholars have promoted the concept of "two-eyed seeing," *etuaptmumk*, which combines Western science with Indigenous knowledge, thus producing an integrative science capable of approaching complex realities in a more sustainable way (Bartlett, Marshall, and Marshall 2012). As Bartlett and her Mi'kmaw collaborators point out, "two-eyed seeing" can also be considered "multiple-eyed seeing," given the variety within both Indigenous and "Western" points of view.

Nepantla Cosmopolitics

The multiplicity of the Moon brings us full circle to the following narration of the Apollo Moon landing by a Zoque ritual specialist, recorded by anthropologist Leopoldo Trejo in Chiapas in 2001:

> When it was announced that the scientists' apparatus was going to arrive on the Moon, Apollo 12 was going to arrive, [but] it didn't arrive. 13 goes, and nothing, 13 didn't arrive. 11 goes, maybe 11 was taken better care of, maybe 11 was more *cabrón*. It arrived and when they went down, things were just like how we are here. But what happened, [the astronauts] wanted to talk, but [those they met] don't talk. [However], two pretty girls, really affectionate—"let's go, let's go"—they did talk, [so the astronauts told them], "Let's go to Earth, it's more beautiful there, let's go. We're leaving on such-and-such a day, we'll come find you here so that we can go see my home, see what the United States is like." They were four days [on the Moon], but they couldn't get around much because there was a river here, a river there, a river everywhere. . . . What's more, the people [on the Moon] didn't talk to them. So, they told the girls they were leaving. "You're leaving, I'll go with you [said one], but you must bring me back."
>
> Those who went to the Moon didn't go straight back to the United States. They fell into the middle of the sea because they were advised that they were going to bring a lot of sickness back with them and they didn't want to contaminate the entire nation. They were picked up in the middle of the sea, in a ship. Their clothes were thrown away because they were contaminated too, and they bathed in the sea. They're arriving in the center of the United States and [one astronaut] says, "Look, I'm really hungry, let's go to a restaurant." "Let's go," says the girl, "let's go."
>
> They get to the restaurant and order their food. What happened was that she doesn't eat, she doesn't even have a soft drink, for two days they had her and she didn't eat, she didn't even drink water. "Well, why aren't you eating?" they asked her. That night, she saw the scientist who took her away from there, she says to him "Look, I gave you permission for two days, no more, don't take any longer, you'll take me back or I'll do what I have to do." She wanted to destroy the United States! [So the astronaut told her]: "Yes, let's go, because [NASA] is scolding me. Eat so that you can hold up," they told her, but she didn't want to . . .

When they got to the Moon, the girl disappeared, they didn't see her anymore, and they felt bad because they didn't know where she had gone and they started to look for her, but they couldn't find her because it's another planet there. We live here, not there. There is the same as here, the store and everything, but there they don't sell what you get here, because the dead are there. (Trejo 2008, 336–37)

Trejo concluded that, although Zoque communities have altered parts of their conception of the cosmos because of their relations with non-Zoque society—the surface of the earth is seen as circular rather than square, seven planetary spheres have taken the place of seven cosmic levels, and the Christian God is referenced as the cosmic creator—aspects of Mesoamerican cosmic logic persist. Instead of the barren, rocky satellite encountered by the Apollo astronauts, Zoques and other Indigenous groups associate the Moon with the underworld, darkness, moistness, sexuality, the feminine, and death. The rivers the astronauts found on the Moon are thus connected to the rivers that the souls of the dead must cross on their way to their postmortem destination. And, like the lunar girl, the dead cannot eat food, only consume essences and aromas. Neither can they communicate with the living face-to-face, only in dreams. The dead can cause illness, however; hence the importance of the astronauts' decontamination after their return to Earth. Trejo hypothesizes that, according to the Zoque worldview, the Apollo astronauts did not travel to an extraterrestrial satellite but a subterranean world that is *other* to the terrestrial world, while bearing an intimate relation to it (344). To me, the narrative also expresses a critique of technoscientific shortsightedness: the one-eyed NASA astronauts went to the Moon believing that the satellite was simply an "eighth continent" (Álvarez 2020), knowable, measurable, explorable, conquerable, never considering that they were traveling to a world that would require a "two-eyed" perspective. The Zoque shaman also slipped in a sly jab at the "powerful" gringos. Through the astronauts' lack of vision, the dead Zoque girl could have ended up destroying the United States!

Some Final Reflections

At one point during his journeys in South America, eighteenth-century naturalist Alexander von Humboldt imagined a future in which humans would not only travel around the globe, but to destinations beyond the confines of Earth. His worried musings seem prophetic: "Mountains of the moon and of Venus! When will we undertake that journey, propagating our culture over other planets—that is, our combination of vices and prejudices—devastating them as the Europeans have depopulated and sacked both Indies?" (in Walls 2009, 234).

I agree with Gustavo Medina Tanco when he argues that a successful Moon mission can be more about the journey than the destination. But I hesitate to accept the idea that the only way to have a voice in decisions about what to do on/with/around the Moon is by sending technology there and converting it to "property." In her defense of "mesopolitics," Isabelle Stengers writes that the "success of a meso device would be to confer upon a situation the power to make those who are attached to it, in an a priori conflictual manner, think together. Not overcome the conflict, but transversalize its terms" (Stengers, Massumi, and Manning 2009). The Moon is an object of attachment for most humans, whether they will ever go, or even want to go there; therefore, activities undertaken on or around the Moon should be subject to an inclusive "thinking together."

In their spatial multidimensionalities, spiral temporalities, relational ontologies, and participatory ethics, Indigenous and artistic lunar worlds offer important alternatives to the dominant technoscientific vision of the Moon as resource or site to be conquered. However, the goal is not simply to replace one lunar world with another, but to develop cosmopolitical negotiations between different ways of knowing and engaging with the Moon. The cases discussed above show that different lunar worlds can productively interact. Scientists and engineers attempt to bridge cultural and technological lunar worlds even as they pursue technological innovation. Legal frameworks struggle to reconcile competing claims on lunar access and meaning. Indigenous communities maintain relational lunar worlds that see the Moon as an active participant in cosmic cycles requiring human care and reciprocity. Finally, artists create spaces where different lunar imaginaries can co-exist, and new possibilities can emerge.

These different lunar worlds are not merely different perspectives on a single Moon, but different modes of worldmaking that generate their own ontological realities. A "two-eyed seeing" nepantla cosmopolitics suggests that, rather than choosing between these worlds, we might learn to see through multiple eyes—Indigenous and scientific, artistic and technological, local and global. This multiplicity of vision could help guide more ethical and sustainable human engagement with the Moon, one that recognizes both its material and symbolic significance across different milieux, "an attunement to the sensibility for diplomacy that cultured nature equipped for finding a commons in the cosmos" (Battaglia 2014).

THE GREAT NORTH AMERICAN ECLIPSE

Seven white satellites
Between my shadow and yours.
Seven moons.

—AURORA REYES, "ECLIPSE OF SEVEN MOONS"

April 7, 2024. Parras de la Fuente, Coahuila. We spent the day in the Chihuahuan desert, which covers large parts of the states of Sonora, Chihuahua, Coahuila, Durango, and Zacatecas. The ecosystem was new to me after spending twenty-five years in southern and central Mexico. However, after our visit to Saltillo's Museum of the Desert yesterday, I felt slightly more knowledgeable about flora and fauna, as well as the prehistory of the region that had been home to various species of Cretaceous-era dinosaurs, in whose demise Mexico also probably played a role when a giant meteor with the Mayan name of Chicxulub struck Earth to the northwest of the peninsula of Yucatán. After a side trip to Mars Station (more on that later), we drove to the dried-up lakebed of Mayrán. When we took off our shoes, I expected to burn my feet, but the ground was cool and slightly spongy because of the water that was still there, underneath the desert sand. Originally, the lake served as the mouth of the Nazas River, but the river was damned up to provide water for Ciudad Lerdo in Durango several decades ago, an event that continues to be a source of tension between the two states. The region is still referred to as La Comarca Lagunera and its inhabitants are laguneros, even in the absence of water. Manuel, one of the organizers of the "KOSMICA Eclipse Expedition," wants us to reflect on the Anthropocene and the land from which we will be watching tomorrow's total eclipse, connecting Earth to cosmic events.

April 8, 2024. We started the day with a ritual of cosmic intention. Manuel had lined everyone up by age outside the house the group had rented on Airbnb, originally the home of the wealthy family of Francisco I. Madero, hero of the Mexican Revolution. We passed through the gate one by one, and Manuel ritually cleansed our bodies with sage smoke. We formed a circle and asked for the goodwill of the spirits of the cardinal directions, the earth, and the sky. We each spoke a word of intention in front of a candle, which would stay lit until our return in the evening, then piled into the two vans that would take us to the Dunes of Bilbao, famous as the setting of controversial filmmaker Alejandro Jodorowsky's trippy *El topo*.

The journey, which usually takes around an hour and a half, lasted longer today thanks to the morning's unusual traffic. We spent the time on the road making an eclipse playlist, after one of our fellow expedition members, Bernard, an astrophysicist retired from the European Space Agency but still very active in the international space community, asked us which songs we would send to the Moon if we had a chance. The van's driver chose Mercedes Sosa's "Gracias a la vida," a banker opted for José Alfredo Jimenez's "Deja que salga la luna," Manuel's mother picked Sinatra's version of "Blue Moon," and Bernard selected "Moon River." People started suggesting other songs inspired by the eclipse: a young artist wanted the Björk song "Cosmogony," a science communicator at the UNAM's Institute of Nuclear Sciences recommended Soundgarden's "Black Hole Sun," a Spanish astrologer was a fan of Sun Ra, and I requested R.E.M.'s "It's the End of the World as We Know It." Eventually, Pink Floyd and David Bowie also made it onto the list. The group's other members—including KOSMICA's director Nahum, a U.S. political economist interested in outer-space policy, a couple from Toronto who worked in tech, an Irish graduate of the humanities program of the International Space University, a graphic design student from California, and, from Mexico City, a radio broadcaster, a psychologist, a systems designer, and a social worker—were in the other van, so they didn't participate in the initial playlist selection.

Families and groups from astronomy clubs and universities had conglomerated near the entrance to the dunes. Like us, most groups identified themselves with eclipse-themed T-shirts and the name of their organization. Fortunately, most umbraphiles who made the trip to Mex-

FIGURE 15 Eclipse. Coahuila, 2025. Photo by Bernard Foing.

ico to see the 2024 total eclipse had chosen other sites to watch the celestial event (like Mazatlán, where NASA scientists, representatives of the AEM, and hundreds of thousands of tourists had gathered), so we found an unoccupied sand hill to set up our cameras and wait, hopefully, for the clouds covering the sky to dissipate. Music blasted from several of the campsites, and one family had inflated a green plastic alien. The members of our group with photography expertise exchanged tips about filters, f-stops, and exposure times. Bernard wanted to scoop NASA and send his pictures to the ESA as soon as possible. Around 11:30, the Moon took its first bite out of the sun, the air grew colder, the light faded, and we started to hear the noise of animals starting to retire to their resting places. We put on our eclipse glasses. One of the participants took a beautiful panoramic photo of the dunes and the darkening sky. An hour later, an opening in the clouds allowed us to see the appearance of Baily's beads caused by light shining through lunar valleys and, finally, a glorious ring of fire around the dark circle of the Moon. I was not the only one with tears in her eyes.

TRANSHABITING MARS

Santo vs the Martian Invasion

The 1967 *Santo vs the Martian Invasion* is a B movie from what many consider to be a low point in Mexican cinema.[1] I watched it one night in October 2023 with about ten other people on the patio of a Mexico City art gallery.

As black-and-white images of rockets, spacecraft, and Earth's cloud cover seen from above filled the screen, a stentorian voice intoned: "As the science of man advances, we are faced with a tremendous question. Is our planet the only one inhabited by rational beings such as us? If not, will we be the ones to conquer those worlds? Or on the contrary, will their inhabitants force us to submit to their domination?" A silver space-ship sped through space, carrying a group of aliens closer and closer to Earth. "From now on," their leader informed them, "we will be speaking in Spanish, the language spoken in the country of our arrival, which the Earthlings call Mexico."

We saw scenes of Mexicans watching television in diverse settings: an upper-class living room, a working-class home, and a cantina. Suddenly, the singing charro on the television screen disappeared, to be replaced by static and then by the image of a group of coldly beautiful men and women with long blond hair, golden lamé suits and capes, and golden helmets that exposed a third eye in the center of their foreheads. "Do not be fooled by our appearance," the leader intoned,

> We are inhabitants of the planet you call Mars. You Earthlings, instead of using your scientific advances for the betterment of humanity, are using them for your own destruction. Thanks to the discovery of nuclear energy and your crazy experiments with the atomic bomb, you are about to upset the planetary system. Before this happens, we must warn you we are willing to disintegrate every inhabitant of the planet Earth.

Three figures watched the broadcast intently: a scientist in a high-tech laboratory surrounded by instruments and flashing lights; a priest in his rectory, the Bible lying open before him forgotten for the time being; and finally, in an office furnished with books and small statues, a bare-chested man wearing a cape, wrestling tights, and a silver mask that covered his entire head and face. This last figure was, of course, Santo el Enmascarado de Plata, the hero of Mexican *lucha libre*.

The Martians recognized that many would doubt their message and, therefore, had decided to use Mexico as a lesson to the rest of the world because, as they would later explain, Mexico had shown itself to be a global beacon of peace by resisting the development of nuclear weapons, and that, perhaps as a result, "its voice might be heard by all other nations." If Earth were to be saved, the invaders insisted that humanity must renounce war forever, establish one global government, and declare one shared language. Santo was visibly perturbed.

The Martian invasion began against the backdrop of a sporting event. One of the aliens appeared suddenly on the field, pointed his "astral eye" at the stands, and "disintegrated" a portion of the crowd. Santo confronted him in the first lucha libre–style combat of the movie, but the Martian used his superior technology to escape when it became clear his opponent was stronger. The Martians decided that "the masked man" was exactly the kind of Earthling they should take back to Mars as the "seed for a new kind of strong and moral humanity." For the rest of the film, they employed various strategies to entrap Santo, from force to seduction, but were unsuccessful.

To resist the invasion, Santo joined forces with the scientist and the priest, both of whom were captured by the Martians. Eventually, Santo found the alien spaceship and rescued the captives in a last wrestling match with the aliens. After leading the humans to safety in the woods, he turned back. "I must destroy that spaceship." The scientist demurred, "No, Santo, please. It could be used by our scientists to unravel the secrets of Mars! Our science would advance five hundred years!" Santo shook his head, "That is precisely what I want to avoid. Humanity, Professor, is not yet ready for such an advance, and you know it." Santo ran back to the spaceship and caused it to explode, barely escaping in time to save himself.

Over fifty years later, members of the viewing public, young, urban sci-fi lovers, had other ideas about technoscience, outer space, and the future, as evidenced by some of the reactions I heard around me at the end of the movie, when the Mexican pop icon blew up the Martians' advanced technology: "NOOOOO!!!!! Fucking Santo! Now I know why Mexico is so screwed! *Por eso estamos como estamos* (that's why things are the way they are)."

Lucha libre, the performance genre that structures all Santo's films, is "a sport in the key of melodrama," in the words of anthropologist Heather Levi (2008, 7), and allows for the "staging of contradictions" in both moral and social terms. The Mexican appropriation of wrestling, imported from the U.S., has been historically linked to the processes of immigration, urbanization, and modernization (212), and has become a symbol of national identity and popular culture. In the last decades, lucha libre iconography has also been incorporated into social movements and popular art. As Levi writes, lucha libre "makes an implicit argument about how social action happens, and what kinds of historical agents are effective in the world" (77). In contrast to U.S. professional wrestling, she argues, in which dramas are usually carried out between actors representing social types, Mexican wrestling "portrays a world in which human agency is limited and supplemented by the intervention of (natural and supernatural) forces beyond human control" and "recognizes that human agency is ultimately constrained by the forces of history, nature, and the world beyond" (78).

Rodolfo Guzmán Huerta began his career in wrestling as a villain ("heel" in U.S. wrestling parlance) or *rudo*; he later joined the opposite band of heroes ("faces") known as *técnicos* or *científicos* (a different form of technoscience than we have seen in earlier chapters), and, borrowing his name from the Robin Hood–like hero of the British TV series based on the novels of Leslie Charteris, transformed himself into "The Saint." On film, Santo faced off against a series of *rudos*, foreign and domestic: vampires, wolf-women, zombies, mummies (both from Egypt and Guanajuato), the Death Brain, the Strangler, the daughter of Frankenstein, the Llorona, and, as we have seen, invaders from Mars. Mexico City chronicler and lucha libre fan Carlos Monsiváis (2009, 131) called Santo "a realist fable of our urban culture" and his films "universal kitsch classics."

But *Santo vs the Martian Invasion* is more than just kitsch. As I have mentioned, 1967 was a pivotal year in Mexico, marking the transmission of the first global satellite program *Our World* and the signing of the Treaty of Tlatelolco prohibiting nuclear weapons in Latin America and the Caribbean. *Santo vs the Martian Invasion* incorporated these portentous events into its plot, as the Martians are goaded into action by the launching of artificial satellites (and perhaps not coincidentally, both films feature singing charros), and decide to land in Mexico in recognition of this nation's stated opposition to nuclear arms. The film concludes with the narrator's voiceover: "Will humanity learn its lesson, or will it persist in its crazy nuclear experiments until it disappears from the face of the earth?"

Looking at the film in its historical context helps to illuminate aspects of how "modernity" and the off-worlding of technology were lived and imagined in Mexico in the latter half of the twentieth century. And, indeed, Santo movies display a marked ambivalence toward technology. On the one hand, his films feature an abundance of mad scientist characters, most of whom are coded as foreign villains. However, as we saw in the trio of protagonists from *Santo vs the Martian Invasion* (wrestler, scientist, priest), Santo often has his own technology, which he tempers with an appeal to faith as a means of combatting the hubris and excessive ambition of those who would abuse science and technology for non-peaceful ends.[2] Again, the rudos are not necessarily villains because they are intrinsically evil, but because they do not play by the rules. Their passions and vices are excessive or unnatural.

On the other hand, the film also highlights Mexico's rejection of foreign influence. David S. Dalton connects, for example, the scene in which the Martians warn Mexico and the world before launching their invasion to the sixteenth-century performance of the *Requerimiento*, a legal document the Spanish conquerors would read out loud to the populations they encountered in the "New World" justifying the Spanish presence through an appeal to natural law and European Catholic justice and demanding the surrender of the Indigenous inhabitants. He further argues that:

> The demand that the earthlings follow enlightened Martian ideals or face extermination equates them with the 1960s United States—a country that appealed to democratic and humanitarian ideals that many Mexicans felt it contradicted in Vietnam. Like Mexico's northern neighbor, the Martians are willing to kill innocent children in defense of their supposed values. (Dalton 2018, 150)

Curator and cultural critic Itala Schmelz (2022, 30) concurs, writing that the B movies of the 1960s that pitted Mexican heroes against alien invaders and other monsters functioned as means of dramatizing the nation's ambivalent reception of the products of U.S. cultural imperialism.[3] Schmelz argues that, as copies-that-are-not-copies of Hollywood productions, these films, and other science-fiction works produced in Mexico, should be understood as examples of what Bolivar de Echeverría termed the "baroque ethos." For Echeverría (2013, 38–39), capitalist modernity is experienced in one of four registers: the "realist ethos" that pragmatically promotes capitalism and denies that any alternative is possible; the "romantic ethos" that sees in capitalism something to be criticized and rejected as opposed to the human spirit; the "classical ethos" that still sees in the capitalist system the possibility of a humanist utopia; and, finally, as we have seen in previous chapters, the "baroque ethos," a cousin to *nepantla*, that parodies and theatricalizes modernity while assimilating it, whose questioning of capitalism is hidden in a series of imitations.

From the Valley of Anáhuac to Valles Marineris

Santo vs the Martian Invasion was a perfect accompaniment to the exhibition *Voyage to Mars: From the Valley of Anáhuac to Valles Marineris*, organized at the Universidad Iberoamericana's Center for Exploration and Critical Thought (CEX) in 2024 by the collective Marsarchive.org.

FIGURE 16 Martenochtitlan and Martelolco. Photo by the author.

This project imagined a Mexican settlement on Mars through the display of objects and images produced by participants in a series of ludic, speculative, and collaborative workshops held in Mexico City between 2020 and 2023. In the words of Marcela Chao, the exhibition's curator and Marsarchive.org's founder, the exposition was to be considered an "expo-fiction" that allowed the public to imagine alternative outer-space futures inspired by urban Mexican collective memories and imaginaries.

Visitors were greeted with a map of two Mars settlements, the twin cities of Martenochtitlan and Martelolco, located inside the canyons of Valles Marineris, in what may have been an ancient lakebed known as

Capri Chasma. Cacao City is drawn over a topographical rendering of the Martian surface copied from NASA sites like Mars Trek and the rendering of Mars in Google Earth, maps which, as Lisa Messeri (2016, 107) has written, seek to make the solar system "democratic, three-dimensional, and dynamic."[4]

Taking seriously this promise of democratic off-earth cartography, the Marsarchive.org map retains some of Mars's official toponyms while transforming and inventing others. Stylized footprints (which represented human movement in pre-Hispanic iconography) march off the map to the east toward Aurorae Chaos, while others head to the west, toward Noctis Labyrinthus. This Mars also has a Cerro del Niño, just north of Ciudad Cacao, and a Puente del Temor that crosses a Muro de Fuego. Eos Chasma, to the south, is the site of the Tomb of Coatlicue and is adjacent to Mictlán. And, in a Martenochca version of "here be dragons," a drawing of a sea monster marks the location of the Mountains of Canauhcoatl, a monstrous boa constrictor. An Aztec-style sun and drawings of Mars's moons Phobos and Deimos grace the map's upper-left corner. The Martian features officially named after Mexican people and places by the International Astronomical Union (IAU)—small towns such as Ocampo, Yelapa, La Paz, Álamos, Taxco, and Izumal; astronomers Francisco Javier Escalante and Elpidio López, who had observed Mars through telescopes; and astrobiologist Rafael Navarro, who had worked on NASA's Curiosity mission—are absent on this map, as they are located far away from Martenochtitlan.[5]

The first room of the exhibition featured the *Codex Marineris*, painted on two long strips of amate bark paper. These narrated the story of a migration of inhabitants of Mexico City to Mars "on a cosmic cactus fruit" in the wake of climate-related disasters on Earth and the pilgrimage they were forced to undertake to establish their city, called Martenochtitlan after the ancient Aztec capital upon which Mexico City was built. The founding event—a "cosmic orgy" that allowed feuding factions of the pilgrims to unite and work together in harmony with the planet—was represented pictorially in a "Martememegrafía," repurposing the popular genre of the *monografía* sold in many neighborhood stationery stores that visually summarize pedagogical information for schoolchildren.

The other side of the room was dedicated to explaining the relation between Earth time and Mars time (a Mars year is 687 days), mediated through the pre-Hispanic Nahua calendar (divided into eighteen months). The combination of the three temporal scales was intentionally disorienting. Hanging in the middle of the room was a Martian globe, which turned out to be a piñata that would be broken on the exhibit's last day, just after the showing of *Santo vs the Martian Invasion*.

The second gallery was devoted to Martelelolco, the city dedicated to "the memory of the place of origin," and featured a series of fanzines drawn by participants in the Marsarchive.org workshop held in Tlatelolco. The third gallery was dominated by a large architectural model (not to scale) of "Cacao City" in Martenochtitlan. The fourth and final gallery, painted a startling yellow, only contained one object: a virtual-reality visor. Visitors who engaged with the visor were submerged inside an orange and yellow geometrical environment created by contemporary artist Ximena Labra as a representation of Mictecacihuatl, a pre-Hispanic goddess of death, in recognition of the traditions that Mexicans might take with them to Mars and the extreme conditions they would face on the red planet.

To understand these cultural objects as more than a kitsch combination of science and culture, "contemporary" and "popular" art inspired by sci-fi imaginings, we must go back (once again) to 2016, to Elon Musk's keynote speech at the IAC in Guadalajara, after a brief detour through the second film that accompanied the exhibition.

Total Recall

Two weeks before the showing of *Santo vs the Martian Invasion*, the same crowd and I had watched a very different cinematic production linking Mexico with Mars, this time in the register of the Hollywood action film.

Arnold Schwarzenegger emerged from a gray train in an underground transportation system. He walked toward a massive building of hulking concrete, with a glass-and-steel entrance slanted into its bulk. Inside, the red-tiled lobby was surrounded by a vertical expanse of rectangular concrete boxes that housed the building's offices. In a computerized directory, Schwarzenegger's character Douglas Quaid looked up the representative of Rekall Incorporated, the brain implant corporation that promised him "the memory of a lifetime," a way to experience exotic travel without ever leaving Earth. Quaid, seduced by recurring dreams of the red planet and a mysterious woman, wanted to go to Mars. In Rekall's offices, whose aesthetics mirror the concrete brutalism of the building's exterior, his adventure began. Two hours later (in movie time), he had traveled to Mars, recovered his buried memories, defeated the evil mining corporation that had enslaved much of Mars's downtrodden mutant population, and discovered the ancient alien technology that would release oxygen into the atmosphere, transforming the planet's toxic environment into a paradise fit for human habitation. Echoing the film's opening dream scene, *Total Recall* ended with a shot of the two protagonists embracing, but now against a backdrop of blue Martian skies. Had Schwarzenegger saved Mars, or were the capitalists still exploiting the planet and its working classes while his character dreamed of a utopian Martian future with a beautiful woman?

The CEX is mere blocks away from the offices of Infonavit, the government-run housing and mortgage agency for Mexican workers that played the part of Rekall in Paul Verhoeven's 1990 sci-fi film. Verhoeven had originally planned to film in Houston, as he felt the city's glass-and-steel architecture would be an appropriate backdrop for a futurist Earth. However, the film's producers insisted that he cut costs, so he took the crew to Mexico instead (Medina 2003, 68). Although initially worried that he would be unable to find suitable filming locations in the five-hundred-year-old capital city, he discovered that the subways and decaying modernist government buildings, built to showcase the nation's mid-twentieth-century pretensions to modernity and materialize the ruling political party's power, allowed him to imagine a different, more dystopian future on Earth and Mars, and all he had to do was "replace a couple of signs" (68). "The present vanishes into the future," wrote Serge Gruzinski, after watching *Total Recall* in Mexico City when it was released. "What a strange feeling to leave the last showing of the film and pass through the labyrinth of the metro, the same passage which, on the screen minutes earlier, propelled us into the next century!" The French historian concluded that Mexico City, whose modernist public buildings co-habit with densely populated working-class neighborhoods, colonial architecture in various states of preservation, and pre-Hispanic ruins, "tends to sow temporal confusion" (Gruzinski 2012, 559).

Not coincidentally, the construction of Mexico City's brutalist architecture required a supply of cheap local labor, a logic that also underlay the choice made by the producers of *Total Recall* to film in Mexico City rather than Houston. "The Mexican film apparatus which, at its peak during the 'golden age' of the 1940s and 1950s served as the Hispanic movie capital, was practically dismantled, and converted into a filmic provider of cheap labor: the maquiladora of Hollywood's dreams" (Medina 2003, 68). The showing of *Total Recall* during the *Voyage to Mars* exhibit somehow manages to condense all the themes discussed here: a futuristic vision of space created in California but resignified in Mexico, the entanglement of Mexico in a global technological supply chain, a national political system that conditions Mexico's participation in the space sector, and a creative, baroque cultural response to the notion that "space is for everyone."

So Far from God, So Close to SpaceX

Is Elon Musk a rudo or a técnico? For Marcela Chao, whom we briefly met at the end of chapter 4, the answer is obvious. In a program about *Total Recall* recorded in 2022 for the Marsarchive.org YouTube channel, Marcela and Mariana Arellano, a young Mexican science fiction author, discussed the character of the antagonist Cohaagen: corporate dictator, owner of the "terbinium" Paradise Mine, violent oppressor of the mutant proletariat, monopolizer of the air supply, suppressor of ancient alien technology, and governor of Mars. "He's who Elon Musk wants to be," they agreed. At the time of the video, Musk had just taken over Twitter (but had not yet ascended to the powerful position he would later hold in the Trump administration). "He wants to silence the diversity of public discourse," opined Arellano, unless it upholds him as "the Messiah of the new Universe." He represents "Gringo Manifest Destiny." "They keep saying that space is for everyone, but for whom, really?"

As I noted in a previous chapter, Elon Musk gave the keynote speech at the 2016 IAC in Guadalajara, "Making Humans a Multiplanetary Species," in which he invoked the planet Mars as "possible . . . something that we can do within our lifetimes."

Musk spoke of his desire to construct a self-sustaining city on Mars and spoke of Mars's closeness to Earth and its geophysical characteristics, as well as the changes that Mars would have to undergo for humans to be able to find water, thicken its atmosphere, plant crops, and build a city. He laid out his strategy for achieving this dream. In 2016, he said, a voyage to Mars would cost ten billion dollars per person. It would be difficult, but costs could be lowered through a series of technological innovations. He explained the technical requirements of rockets and ships, fuel composition, and the infrastructure that would be necessary to maintain a Mars system. At the end of the presentation, the digital planet on the screen behind him revolved, transforming itself from a red desert sphere to a red, blue, and green sphere with an atmospheric halo indicating its habitability. This last image, a disquieting hybrid of Mars and Earth, accompanied Musk for the rest of his speech.

Marcela Chao saw the conference on YouTube and was captivated by Musk's passion, if not by his plans or personality. In 2016, Chao was working in Mexico City at the Center of the Image on projects that combined science, technology, and art. She founded the organization that would later become Marsarchive.org with her friends and friends-of-friends as a way of intervening artistically in the photographs generated by NASA's Mars rovers, but the project eventually expanded to include reflections on "the reactivated space race" led by space agencies and the private sector. Just in that moment, Elon Musk came to Mexico and, as Marcela recounted to me, "I got archive fever."[6] She decided that Marsarchive.org would compile all existing information about Mars, like a contemporary version of the General Archive of the Indies in Seville, to "generate accountability" because everything is changing so fast in the space industry and, "how long before the millionaire gets bored, right?" She contacted various space actors in Mexico, including the AEM, the Center for Digital Culture (CDC), and the SGAC, and Marsarchive.org began to grow and expand. An interdisciplinary group formed around Chao made up of mostly millennial-generation students and professionals in, among other fields, art, design, literature, engineering, and astronomy.

Between 2017 and 2018, Marsarchive.org presented events such as the curatorial program MartePop, a Wikipedia edit-a-thon, a series of projections of movies about Mars, and a virtual-reality exhibition. They continued with the first edition of the Christmas *Posadas marcianas*, a series of podcasts, and the festival *La bendita primavera marciana* ("Blessed Martian Spring") during the annual celebrations of Yuri's Night that commemorate Yuri Gagarin's voyage to space in 1961. Some members of the collective (many of the SGAC affiliates) had more "scientific" perspectives and decided to "go down another road," while within the group, according to Marcela, "the craziness kept going like we wanted it to."

"Why Mars?" On the group's website, Marsarchive.org answers the question they hear the most with an explanation that acknowledges the presence of Elon Musk in the discourse of outer-space exploration without ever mentioning him by name:

> Becoming a multiplanetary species will depend on a series of scientific, technological, economic, political, and logistical factors, as well as social and ethical ones, in which environmental and decolonial perspectives cannot be ignored. Turning our gaze to Mars invites a profound reflection on our desires, goals, and ways of acting, moving toward more benign ways of relating to each other and to the space we inhabit.[7]

Martenochtitlan

Martenochtitlan was born out of a discussion between Marcela and Juan Claudio about the possibility of founding a Mexican city on the Mars landing site of the rover Curiosity and was visualized as a "Neo-Tenochca futurist meme." They decided that the meme deserved further attention and began to develop a workshop that would take up the possibilities of a Mexican inhabitation of Mars. Would the first Mexican Martian city be built on a lakebed periodically shaken by earthquakes, as was Tenochtitlan and, later, Mexico City? And, thinking about Elon Musk's plans for Mars, do we terraform Mars or do we transform ourselves?

In 2020, Marcela and Juan Claudio met up with their friend Amadís Ross, a researcher at the National Institute of Fine Art and Literature (INBA) and expert in science fiction. Until then, Marcela told me, everything they did concerning Mars was "universal," but "really when I say universal, well, everything was very American, you know?" For Marcela, this "universal" perspective would change through her participation in the seminar "Aesthetics of Science Fiction," organized by the INBA's National Center for Plastic Arts Research, Documentation, and Information (CENIDAP), and coordinated by Amadís, who was convinced of the importance of a decolonial view, and, as he emphasized, the need to "create science fiction from where you are." According to Marcela, this perspective was fundamental because Marsarchive.org was meant to imagine futures somewhere else, "although this future wasn't realist, I mean, 100 percent scientific." Amadís was inspired by the name Marcela and Juan Claudio had given to their imaginary city because "Tenochtitlan has great power in the Mexican imaginary . . . it touches the heart of Mexicans." The group is aware of the problems generated by, first, appropriating precolonial imagery, and, second, focusing on the center of Mexico and the Aztec image to the exclusion of other pre-Hispanic populations: it could have been, they say, "Marte-Chichén Itzá." But, as Amadís insisted, other names wouldn't have the same mythical or sonic force. "Martenochtitlan" opens the door for an appropriation of the mythic symbolism of the nation's center, allowing for "dialoguing with the universe" and imagining new futures.

They decided to present the workshop in three parts: in the first, participants would work with the mythic aspects of the narrative to be produced; in the second, they would develop the architecture and urban planning of the imagined city; and finally, they would write up a social contract to guide the Martenochcas in their new civilization. However, by the end of 2020, they had only been able to fully develop the first part of the workshop, which was sponsored by the UNAM's Program for Art, Science and Technology (ACT). Around sixty people responded to the call for participants, and twenty were selected considering applicants' diverse backgrounds in terms of discipline, gender, and age, although the majority were urban, educated, and middle-class. Fifteen people ended up in the workshop, where they heard Juan Claudio talk about "the reality of Mars," emphasizing the planet's inhospitable nature, difficult terrain, dust, radiation, and, above all, its lack of easily available water. "Everything you need to know if you're going to live on Mars," he said, so that the myth constructed by the participants would contain at least some reference to physical and technical realities.

Workshop participants also read the myth of the founding of Tenochtitlan, as well as other literary texts, including *Invisible Cities* by Italo
Calvino (1974), Ray Bradbury's (1950) *The Martian Chronicles*, and Kim
Stanley Robinson's (1992–1996) Mars trilogy. They reviewed information about the planet's topography so that they could plan out possible
routes, obstacles, and settlements, with the goal of writing a mythic narrative and illustrating a codex that would offer an alternative vision of a
human future on Mars, imagined and produced in Mexico.

The first decision the group had to make was how to respond to the
question posed by the organizers: "Do we want to 'colonize' Mars?" The
(almost) unanimous answer was "absolutely not." The collective feeling
went against what Juan Carlos characterized as "what is happening in
space discourse all over the world, especially in the United States, becoming more neoliberal than ever." Workshop participants decided that
the narrative of the Tenochca migration to Mars should show the Martenochcas "co-habiting, learning to live with the planet," without falling
into colonialism or militarism.

> Before our times, when we only lived on the Old Island, little by little our
> ancestors were making Mother Earth, and themselves, sick. They had lost
> the divine *elhuayotl*, the soul that connects us with the world. Then the
> gods Caktikaktéotl, god of the void, Ehécatl, god of the wind, and Teoh-
> yotica Xalli, goddess of dust, came to us through portentous signs and
> charged us with a great mission: find a new land in a far-away place, where
> we could make communion with them and recover the *elhuayotl*. In the
> new land, the chosen founded Martenochtitlan, a great city of reflection
> and purification. During this voluntary exile, the travelers would fight the
> demons of the world as eagle warriors, to rediscover their eternal essence.
> Through misery and sacrifice, distance, and nostalgia, they would cleanse
> their hearts and then return to the Old Island, carrying their message of
> peace and joy to its inhabitants. The new world would save the old one
> from itself. Mother Earth would be able to breathe again.[8]

The organizers divided the participants into three groups to work on
each "act" of the myth. In the act called "The Voyage," the first group
would imagine the immigrants' journey from Earth, the Old Island, and
arrival on Mars, the Red Island. Participants decided that, after hearing

the call of their gods, "the great lords" would need to build a radically different ship than Elon Musk's Dragon, with its mythic European nomenclature and phallic structure. The future Martenochcas created "The Great Quetzalcóatl, the leaden plumed serpent, an enormous ship with golden metallic plumage that would cross the obsidian rivers that flow between the cosmic islands."[9]

The ship would make three voyages: the first two crewed by robots whose task was to find water and build the foundation of the imagined city, and the third would be the beginning of the human settlement of Mars. After the first trip, the robots informed "their masters" that they had found frozen water in the Hellas Basin, and so the humans undertook their voyage. "The great lords were sure of the power of their science and their ingenuity. Their arrogance would follow them to the new land." Four hundred passengers (the *Centzontli*, twenty groups of twenty, a sacred count in Mesoamerica and the basis of precolonial Nahuas' numerical system) were chosen to undertake the voyage to the Red Island. Despite their careful planning, when they arrived, they realized that there was not sufficient water under Hellas's surface. Therefore, the Centzontli were forced to search for a new place to establish their city. They sent the machines to the four cardinal directions to determine the most propitious site.

Group number two worked on this phase of the narrative titled "The Pilgrimage." They wrote that, of the four machine-led expeditions, "three were lost in the dust storms, and nothing was ever heard from them again." The bravest of them traveled on the surface so they could scout ahead. To protect themselves, they used suits made of nopal and communicated with positronic devices carried by designated *tameme*, or porters. Eventually, the communicators acquired an "almost divine voice."

During the voyage, the pilgrims encountered new life-forms: small bacteria with which they underwent "a kind of fusion or *mestizaje*." This fusion caused a sickness that affected their mucus membranes and bones, transforming human bodies into "marzipan statues that ended up falling apart." The Martenochcas interpreted the disease, which resulted in the death of half their population, as a punishment from the gods. Between this and other calamities, twenty-five years passed. "Some people died, others were born and grew up; these were much better adapted to the conditions of the Red Island."

Eventually, the pilgrims arrived at Gale Crater, where they came across the ruins of SpaceX City, a settlement that had been founded by "the multibillionaire Elon Musk and his followers." All its inhabitants had died of the disease caused by the Martian bacteria.

> After much journeying, the decimated Centzontli arrived at Gale Crater, where they found the settlement known as SpaceX City. Worn out, almost dead from hunger and thirst, they found rest and foodstuffs, containers of water, and an ansible, a device that permitted simultaneous communication with the Old Island. The Centzontli informed their site of origin that death ruled in this place: all the people sent by the multibillionaire Elon Musk to inhabit the city were dead and dried up.

"Perhaps the bacteria brought out what the inhabitants carried with them: colonialist ideas, war, resource expropriation, xenophobia, and misogyny." Fortunately, they encountered in the ruins an ansible—an apparatus that made possible communication at speeds faster than light that had been imagined by the U.S. writer Ursula K. Le Guin in 1966—that let them know that there were natural resources available in the Plain of Tharsis. (As Marcela said to me, "In the end, they couldn't completely let go of foreign science fiction.") The Martenochcas continued their pilgrimage for another fifty-two years, a sacred Mesoamerican cycle.

The third group of participants worked on "The Founding," the narrative's conclusion. Inspired by a collective dream of Teohyotica Xalli, the goddess of dust, the pilgrims went to a cave, which they entered, and thus began the last stage of their journey. The cave was dark, and the pilgrims were scared and uncertain. But a little girl spoke with the voice of the goddess and motivated them to continue. They found an ancient underground city, where they obtained fundamental knowledge of the planet's nature. After emerging from the cave, they crossed the rugged landscape of the Labyrinth of Noctis and finally arrived at their ultimate destination: Valles Marineris, where they would build Martenochtitlan. "In honor of their origins, and as a means of uniting the place from which they had come and the place they had found, they erected a blue city, which contrasted with the red of Mars, a city in harmony with its planet. It was clear to the Martenochcas that they and the Red Island were one. There was no place for arrogance on this planet."

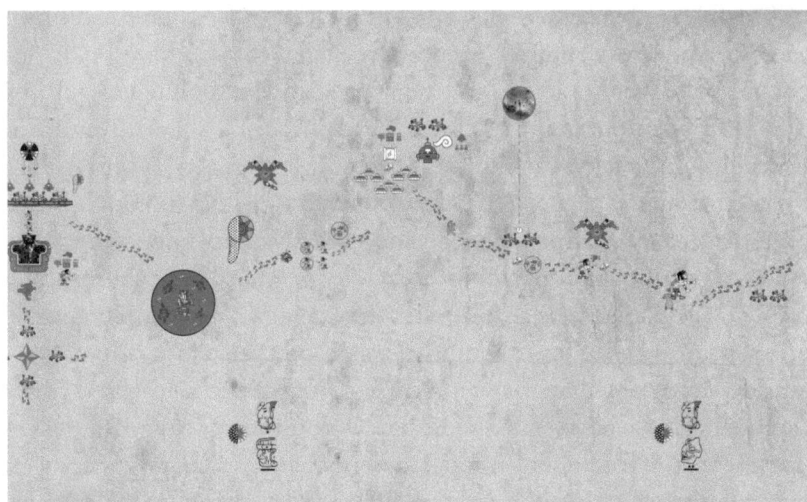

FIGURE 17 *Codex Martenochtitlan.* Courtesy of Marsarchive.org.

In the last session of the workshop, the participants drew a codex that would illustrate the myth they had created. Working from a collection of Mesoamerican glyphs that Amadís had provided them, participants chose the symbols that they deemed appropriate. In some cases, they created new icons, or fused icons, to represent futurist notions such as the spaceship, but they maintained the aesthetic of their source material. Although, as Amadís stated, they had the option to download internet images, they decided not to "contaminate" the original iconography.

The second edition of the workshop, "Martenochtitlan: Myth, Rite, and Site," expanded on its predecessor's themes. Again, participants— mainly students and young professionals, whose backgrounds included art, architecture, literature, theater, psychology, graphic design, finance, computer programming, engineering, and chemistry—reflected upon pre-Hispanic myths and decolonial science fiction, learned about Mars's environment, and worked in groups to create a migration narrative and iconographic codex. Again, the question was posed, "What kind of world will we imagine?"

"We'll leave because we have to," said Quino. "Not just because we want to explore or colonize." Sandra agreed that colonization should not be the goal: "We should be like [the naturalist] Humboldt, and not Columbus."[10] The story began with an apocalyptic vision of Earth's future in which the planet had become desertified or "Martianized" as the result of the cataclysmic impact of a cosmic ray that was an ultimate consequence of human-caused climate change and excessive dependence on technology. Global communications broke down, and society collapsed, in parallel with Indigenous society after the Spanish conquest. But groups of Indigenous and poor urban Mexicans from the suburb of Ecatepec, already "accustomed to crisis and scarcity," organized themselves to take care of the few available natural resources.[11] At the same time, there was a resurgence of traditional practice, and Indigenous shamans once again became important sources of knowledge about Earth and the cosmos. During nightly meetings, these leaders allowed people to once again "recognize themselves in the stars, and remember that, before electricity lit up the night sky, there were stars, nature, and spirit." Rather than technology, cultural diversity and cultural mestizaje became sources of strength and hope for the future.

Everything changed when a meteor fell to Earth, landing on the summit of a mountain in Mexico City. Eventually, it became clear that the stone was a message from the planet Mars, calling for an exodus from Earth, as the planet "needed a rest, needed a rest from us." A seed came forth from the meteorite and grew into a giant nopal plant capable of space flight. And between mystical trance and advanced technology, the Ehecatelpeños (a combination of Ecatepec and the name of the wind deity Ehécatl) boarded the nopal ship and traveled to Mars, where they landed in the Labyrinth of Noctis and began a new era in their history.

Like the Martenochcas from the first workshop, these Earth exiles also found a settlement, "with certain characteristics of a private corporation." They encountered humanoid beings, Martian representatives of the corporation, who wanted to enslave them and subject them to genetic experiments. The Ehecatelpeños escaped, along with a "mestizo" Martian-Earthling hybrid, who betrayed them to the Martians. A series of battles, in which the Martians were guarded by the nocturnal gods Phobos and Deimos (personifications of the two Martian moons) resulted in the death of many Ehecatelpeños. However, the mestizo came across a sacred cactus-like plant, ate its fruits, and was transformed into an avatar of Huitzilopochtli, the Aztec god of war, becoming the protector of the Mexican immigrants. After a final bloody battle in which both Deimos and Huitzilopochtli perished, the Martians were defeated and returned to their city with the body of their fallen protector. Phobos decided to align herself with the Ehecatelpeños and was transformed into the Mexica goddess Coatlicue, whom we saw in the first chapter.

The narrative then described a period of conflict between traditionalists who called themselves "Rubrum" and were tied both to Mars and their ancestral beliefs, and "Technos" (note the reference to lucha libre's rudos and técnicos) who believed in the power of metal and technology to lengthen their human lives and thus avoid as long as possible their transformation into Martians. The Rubrum delved into the planet and constructed underground dwellings, while the technohumans were exposed to cosmic storms and affected by a fatal disease known as "the red fever." Eventually, the two sides negotiated a peace treaty. They exchanged technology and culture, re-establishing their city in the Plain of Argyre, adapting themselves "more ethically and efficiently" to their surroundings, and creating new rituals to celebrate their union.

As well as a myth and a codex, participants in the second edition of the workshop created multimedia mock-ups of Martenochca architecture and descriptions of commemorative rituals accompanied by music playlists. They also repurposed the popular "meme-graph" genre that combines internet memes with Mexican current events and cultural phenomena, constituting in its turn a riff on one of the most traditional representational forms of primary education in Mexico—the "monograph," a letter-sized infographic sheet about diverse topics with a series of illustrations on one side and textual information on the other. The Martememegrafía condenses the group's ideas about the main Martenochca celebration that commemorates the major historical experiences, from their arrival in the nopal spaceship to their conflicts with the native Martians, to their civil war and the final re-establishment of Martenochtitlan. These events are illustrated with a collection of collages combining the iconography of the *códices* created for the workshop with historical and popular cultural images.

Martenochca narratives and worlds are messy, cobbled-together ideas that do not hide the organic, collective process through which they were constructed. This is design as world-building, in the register of the gift rather than the commodity. Martenochtitlan continues what Schmelz (2022, 44) calls the B-film tradition of mixing high tech with Indigenous iconography, a juxtaposition that produces a playfully uncanny combination of recognition and estrangement: homely Mexican icons, places, and referents inhabit the radically unfamiliar Martian territory. Well-known markers of national identity, like pre-Hispanic deities and native plants, eventually allow the Martenochcas to adapt to a new, initially hostile environment. Nopales, maize, and *cempasúchil* are transformed into technological devices that appropriate "Indigenous scientific literacies," to use the phrase that Grace Dillon (2012, 7) mobilizes to characterize "those practices used by Indigenous native peoples to manipulate the natural environment in order to improve existence in areas including medicine, agriculture, and sustainability" and that "[stand] in contrast to more invasive (and potentially destructive) western scientific method." Martenochtitlan also allows technological mechanisms like artificial intelligence to co-exist with divine forces, whose will and favor also influence events. The colony's eventual success in contrast to the destruction of Elon Musk's SpaceX City results from two factors absent in most discourse about the future terraforming of Mars: the need to adapt to the planet instead of forcing the planet to adapt, and the value of cultural memory as a resource for survival.

"We aren't Mexicas anymore, but Mexic-anos," Amadís told me. The *raza cósmica*–inspired ideology of mestizaje became the ultimate technology for survival on Mars, but instead of mixing Indigenous and European characteristics Martenochca mestizaje mixed Indigenous-inflected *mexicanidad*, contemporary global cultural references, and Martian ecology. Ursula K. Le Guin's ansible, the spaceship Quetzalcóatl, the cactus suits, the energetic beverage similar to pulque, the artificial intelligence apparatus carried on the backs of the *tameme*, and others became devices for transhabitation, a process that Valerie Olson (2018, 151) writes of as the design practice of creating habitats on Earth that will or could be inhabited off Earth. Olson's conceptualization highlights the paradoxical nature of the work of transhabitation. Designing outer-space habitats is an elite practice undertaken in the exclusive settings of the Global North institutions promoting the human exploration of outer space; however, the designers working on these projects are often heavily invested in the transgressive, transitional (the prefix "trans" is key) potential of architectural design that will make collective, ecologically sustainable existence possible in extreme extraterrestrial conditions (148). Martenochca transhabitation is explicitly anti-elite, pointing to the kind of social design that Arturo Escobar (2018, 25) characterizes as a fundamental human capacity not limited to the professional designer. The project's recognition of the trauma of migration and of the globalized nature of cultural transmission also adds the idea of the "transnational" to the repertoire of transitional processes indexed in the practice of transhabitation (Olson 2018).[12]

Martelolco

The plans for the gallery dedicated to Martelolco in *Voyage to Mars: From the Valley of Anáhuac to Valles Marineris* originally included a three-dimensional *puesto en abismo*, or mise en abyme, a typically baroque heterochronotopic object that would allow the public to look through a series of stratigraphic or archaeological deposits, giving viewers a sense of the regressive density of time underlying one of Mexico City's iconic sites. It was to have been one of the products of a workshop held in an urban orchard in Tlatelolco that was dedicated to the imagining of Martenochtitlan's companion city, Martelolco, Unfortunately, the piece could not be finished in time, although the idea of Tlatelolco as a site that is intimately entangled with layers of collective memory was expressed in other materials that did form part of the exhibition.

The city of Tlatelolco was founded in 1337 by a group of Mexica nobles who separated from the Tenochcas as a response to their dissatisfaction with the political regime. The two cities were alternately rivals and allies until 1473 when the rulers of Tenochtitlan conquered Texcoco and incorporated Tlatelolco, coveting its market, which was one of the wonders of the pre-Hispanic world.[13] The conquistador Bernal Díaz de Castillo dedicated several pages of his chronicle *True History of the Conquest of New Spain* to a description of the market that seemed to him to be "much wider and larger than that of Salamanca. . . . [W]e had not seen such a thing, we were astonished at the multitude of people and quantity of merchandise and at the good order and control they had everywhere" (Díaz del Castillo 2012, 404). Thirty pages later, Díaz de Castillo wrote of another scene of excess, occurring in the aftermath of the final defeat of the Mexicas, two years after he and his fellow Spaniards first marveled at the market of Tlatelolco, now the site of the capture of Cuauhtémoc, the last emperor of the Mexica:

> All the houses and fighting platforms in the lake were filled with heads and dead bodies, and I do not know how to write about it, for in the streets and the very courtyards of Tlatelolco, it was all the same, and we could not walk except among bodies and heads of dead Indians. (434)

Using its scattered and demolished stones, the Spaniards promptly erected a church on the ruins of Tlatelolco. The Temple of Santiago (named for Hernán Cortés's patron saint) still stands today, representing the colonial aspect of the Plaza of the Three Cultures. The great pre-Hispanic market was abandoned.[14] While Tenochtitlan (rebaptized San Juan Tenochtitlan) became Mexico, and the capital of New Spain, Tlatelolco was designated the capital of the *República de Indios*. The Franciscan Imperial College of the Holy Cross of Tlatelolco was founded in 1536 as the first institution of higher learning in the Americas, where

Nahua medicine, Christian theology, Spanish, Latin, the humanities, and political science were taught to the children of the caciques, Indigenous leaders who would govern the colony's native settlements. Works documenting pre-Hispanic knowledge of the natural world, such as the *Badianus Manuscript* (the first Indigenous herbal in America) and the *Florentine Codex* (a register of Nahua history and science), as well as texts by the Spanish evangelists like much of Sahagun's encyclopedic *General History of the Things of New Spain*, were produced in the College of the Holy Cross. Here can be seen the seeds of Vasconcelos's dream of a "cosmic race." However, a combination of epidemics, economic woes, and opposition to the education of the Indigenous elites resulted in the college's abandonment in 1576.

During the colony and posterior periods, Tlatelolco and its surroundings fell into decay. The extensive landfilled areas of Lake Texcoco surrounding Tlatelolco were used as collective gravesites for plague victims and, later, casualties of the Mexican Revolution. During the late nineteenth century, much of the land passed into the hands of private companies that built warehouses and railway depots, displacing the local population. A military prison famous for housing political dissidents was built on the spot once occupied by the College of the Holy Cross, and what had been the seat of Indigenous authority in Tlatelolco was turned into a correctional school.

In the 1960s, modernist architect Mario Pani was tasked with the planning and design of the ambitious urban housing project to be built on the site of the ancient market of Tlatelolco, after demolishing the industrial wasteland that had grown up in the past century: the Nonoalco Tlatelolco Urban Complex, which would be the largest apartment development in Mexico, occupying nearly thirteen million square feet and consisting of 130 buildings. The urban renewal project was intended to materialize what Carlos Monsiváis (1988, 54) termed the "modest utopia of a Mexico without tenements." The original project was designed to accommodate seventy thousand inhabitants, including the seven thousand who would be displaced by the construction, although these were ultimately unable to pay for the apartments that had been set aside for them.

There were to be three sections, referencing the typical periodization of national history: Independence, Reform, and Republic. Each building, named for a Mexican state, would house three types of residents according to their socioeconomic status: *A* for the working class, *B* for the middle class, and *C* for the affluent. The project was imagined, following the model of Le Corbusier, as a city within a city, and would include all necessary services, commercial venues, schools, clinics, social clubs, and green and recreational areas. Only foot traffic would be permitted, and the daily "association" of its inhabitants of all classes was encouraged by the complex's design. Appropriately, given the context of the Cold War and the Space Race, one of the playgrounds meant to support the healthy growth and education of the children inhabiting the apartments was designed in the shape of a forty-foot-high rocket, notably modeled after Soviet space technology (Solano 2018, 115). Although cultural critic Rubén Gallo compares the complex to the construction of Brasilia four years earlier, in the context of this book, the whole project could be considered in hindsight a kind of space colony prototype. "All was mathematically calculated: the size of the apartment a middle-class family would require, the parking spots, the distance between residences and businesses, the square meters of garden each inhabitant would need. The apartment towers were 'habitation machines' in which everything—water, gas, sewage, ventilation—could be calculated and distributed in the most efficient and ordered way" (Gallo 2020).

In recognition of the area's history, Pani's large-scale architectural project was accompanied by a large-scale archaeological project that continued and expanded on the excavations that had been carried out throughout the twentieth century. The center of the complex was to become the "Plaza of the Three Cultures," surrounded by the excavated ruins as evidence of the pre-Hispanic period, while the Church of Santiago, the Convent of San Buenaventura and San Juan Capistrano, and the College of the Holy Cross would represent the colonial period. The third culture, imagined as modern Mexico, would be represented by Pani's housing project, as well as the Tlatelolco Tower, designed by Pedro Ramírez Vázquez, who had also been one of the architects of the National Museum of Anthropology, inaugurated in 1963. The tower would be the seat of the Secretary of Foreign Relations (SRE), which served as the site for the 1967 signing of the Treaty of Tlatelolco, which was, as we have seen, the implicit motivation for the Martians' decision to start their invasion of Earth with Mexico in the film *Santo vs the Martian Invasion*. As the symbol of Mexican economic and technological modernity allied with its cultural and historical identity, the plaza would feature heavily in the 1967 program *Our World*.

However, as we have seen in previous chapters, both the dream of Mexico as a leader in the struggle for world peace and Pani's dream of a socially and ecologically harmonious housing project were abruptly shattered only a year after the signing of the Treaty of Tlatelolco, when on October 2, 1968, the Plaza of the Three Cultures became the scene of a brutal massacre of student protesters at the hands of the Mexican government headed by President Gustavo Díaz Ordaz. Ironically, "as the military opened fire on the students, Nonoalco-Tlatelolco's 'typically modernist elements' left the students particularly vulnerable: few access points, sidewalks designed to lead to the Plaza of the Three Cultures and a group of towers that created a panopticon—which allowed the army to corral and execute the students. . . . Modern architecture is also an architecture of control" (Gallo 2020).

The contrast between a show of international pacifism and the internal violence exposed by the government's oppression of social movements was to reveal the cracks in the hegemony of the ruling Institutional Revolutionary Party. The party's brutalist architectural aesthetic was also dealt a death blow. As Castañeda (2010, 119) writes, "That the massacre took place in the plaza has had a lasting effect in determining the critical afterlife of all the architectural spaces of 1960s Mexico." Toward the end of his life, Mario Pani complained that, instead of the pinnacle of modern urbanism to which he had aspired, many Mexicans thought of the complex he designed as "the place where they kill students" (Gallo 2020). Indeed, as the slogan "October 2 shall not be forgotten" suggests, the massacre has overshadowed the precolonial Tlatelolco market, the 1967 Treaty of Tlatelolco, and the 1968 Olympics in the collective memory. Rather than achieving its goal of becoming an architectural expression of a unified Mexican identity that incorporates disparate historical moments and cultural elements, The Plaza of the Three Cultures became a palimpsest of violence. But it also became, for some, the cradle of modern democracy, as reflections on the brutal repression of the student movement paved the way for an eventual opening of Mexican society and politics (Allier-Montaño 2015, 131).

Today, around twenty-seven thousand people live in the Tlatelolco complex, although many of the buildings have deteriorated in the last decades. The famous Rocket of Tlatelolco became increasingly dangerous, and by the end of the 1970s it was used as a refuge by the area's unhoused population. It was later demolished. The Nuevo León residential building collapsed during the catastrophic earthquake of 1985, which also resulted in the demolishing of nineteen other structures. More damage occurred in the wake of another earthquake in 1993. The Tower of Tlatelolco was acquired by the UNAM and converted into a museum of the area's history and culture. A monolith erected in 1993 and carved with the names of twenty of the victims of the 1968 massacre stands in front of the chapel of Santiago, and a giant sundial marks the site of the collapsed building.[15]

But none of my Space Generation interlocutors were alive in 1968. The young poet Víctor Joel Armenta gives voice to their understanding of Tlatelolco: "I didn't see death / in these ruins that I now see . . . I didn't see the lightning flare / the bolt of death falling from the sky. / I did not see in Tlatelolco that stabbing death. . . . We are a different generation that circulates in the infinite veins / new people / ripping experiences from the memory of the others" (in Sanchis Amat 2020, 31).

The Tlatelolco Urban Orchard is located on the site of the demolished Oaxaca building a few blocks away from the Plaza of the Three Cultures. This was the meeting place for the workshop "Martelolco: Portable Memories for a Future on Mars." In the fanzine that would be one of the products of the workshop, organizers Marcela Chao and Alejandra Espino explicitly invoked the French historian Pierre Nora's "sites of memory," or *lieux de memoire*, as "places where memory crystallizes and takes refuge," and "those places in which the worn-out capital of collective memory are anchored, condensed, and expressed" (Marsarchive.org 2023). The weekly encounters were to be both exercises in remembering and a critique of the loss of memory. Throughout the five sessions held during summer 2023, participants would engage each of their senses to anchor their memories or imaginaries of Tlatelolco and decide which ones they would take with them on a voyage to Mars.

What would Mars taste like? "It would taste insipid, with a touch of blood and salt," said the organizers. What would Mars smell like? "Acrid, like chalk and rusted metal." Participants created "Martian" recipes inspired by the orchard's culinary offerings, which were vegetarian and made with ingredients harvested on-site. But "Martian beans" would likely take much longer, as they would have to be transported from Earth. What would Mars feel like? "Cold, dry, and rocky," was the answer. Participants made rubbings of different objects and structures found in the orchard to evoke the textures of memory. What would Mars sound like? "Not much," they were told, "because the Martian atmosphere can't transmit sounds like the one on Earth. It would be strange and silent." Participants wrote Martian-inspired music.

When I visited the orchard during the last session of the workshop, dedicated to the sense of sight, I encountered frogs, fish, algae, trees, bushes and flowers, ponds, rows of plants waiting to be harvested, a dog, and several cats. A painted wooden sign featured a flying saucer beaming up a flower, with the warning "Don't abduct the plants!" What does Mars look like? The organizers responded that sight is the easiest sense to apply to a perception of Mars because we have the photos taken by rovers sent by NASA and other space agencies. There is less light on Mars than there is on Earth, and the suspended dust tints the photos orange. When the sun sets, it has a kind of blue halo. Just as we can sometimes see Mars from Earth, we can sometimes look back and see Earth from Mars. After the initial discussion, we went off to draw things we found visually interesting in the orchard. I returned with an unrecognizable chalk drawing of stones and water lilies, while the others had drawn leaves, flowers, and animals with much more artistic ability, evoking the more-than-human vibrancy of this small oasis encircled by concrete in the center of Mexico City. Throughout the workshop, participants were asked: What would you take with you to Mars? What would you miss most? "The food, trees, the ocean, pets . . ." I don't think a single person would have signed up enthusiastically for a one-way ticket to the red planet.

The sensorial exercises undertaken by the workshop's participants would be collected into personal fanzines, or "portable memories" in the words of the organizers. Many of these found their way into the exhibition *Voyage to Mars*, where they were hung from string along the walls of the gallery dedicated to Martelolco, the Martian city of memory. In a nod to Mars's radically different planetary sensoryscape, a label on the wall at the entrance to the gallery stated that the fanzines were to be understood as a series of documents brought to Mars from Earth "to preserve the memory of the ways in which humans sensorily perceived their environment." Mexican visitors would comprehend the link between Tlatelolco and memory without the need for further explanation. And some would have made another connection, between the speculative experience of immigrants to Mars and the remembered or narrated experience of immigrants to the United States. I heard one visitor exclaim, "It's true, the beans are not the same!"

In May 2024, in the middle of a weeks-long heat wave in Mexico City, I attended a second workshop in the urban orchard called "Worldbuilding Martelolco." Most of the participants were repeaters, core members of Marsarchive.org and faithful contributors to its WhatsApp group chat, exchanging Martian memes and news, party and karaoke invitations, media recommendations, and pop-culture gossip. But there were some newcomers, as well: a French artist doing a residency on art and outer space in Mexico, my daughter visiting from Puebla, and a mother from the neighborhood who brought her two small sons. The first session was dedicated to a collective reflection on why humans might leave Earth. We talked about the orchard, scales of life, definitions of life, and the Gaia hypothesis. On butcher paper taped to the table, we drew how we imagined Mars and what we might miss there. The second session was dedicated to a talk by Juan Claudio (for the newbies) to try and connect our imaginations with Martian reality.

During the third session, despite the heat, we took a walking tour through the "utopic city" of Tlatelolco. Fermin, a Marsarchive.org contributor who lives in the housing development, was our guide. We walked through the Plaza of the Three Cultures, weaving around the iconic structures that had been built centuries apart, as well as the monuments to the memory of the massacre and the natural disasters that followed. We wandered around the archeological site, then passed through what was left of the colonial *tecpan* (royal house) built to control the market's commerce, today decorated with Diego Rivera's mural of the conquest, *Cuauhtémoc Against the Myth*, in which he pits the Spanish horse against the Mexica plumed serpent Quetzalcóatl. We caught a glimpse of Juan Diego's baptismal font in the chapel, but we couldn't get close because a quinceañera was celebrating her fifteenth birthday with a mass.

We took pictures of decades of mural art on the walls of the housing development's towers and the unmistakable Chihuahua building where the massacre took place. Fermin explained the ingenious carts that traverse the covered walkways to pick up garbage from the buildings (since no cars are allowed). I saw a flyer for the Martelolco workshop pasted to a wall, a space it shared with graffiti and advertisements for a masseuse, a home-repair service, and an organic product for getting rid of unwanted insects. I tried to imagine the pre-Hispanic Tlatelolco market or the colonial capital of the Repúblico de Indios, although the intensity of spatial and temporal palimpsests, as well as the knowledge that I was walking over what had been the site of a brutal massacre, finally overwhelmed my capacity for mentally inhabiting ancient history. At any rate, the plaza continued to be transited by people making new memories on that Saturday: resident families taking a stroll, people waiting on the sidewalk for the Metrobus that traverses the city from north to south, tourists taking pictures, and vendors hawking a variety of products including commemorative objects with the slogan "October 2 shall never be forgotten."

The final session of the workshop was dedicated to the creation of the city Martelolco. Participants were divided into two teams, one of which was instructed to think about a Martian world established by nation-states, and the other asked to imagine a world built by private enterprise. The first group ended up with a city populated by convicts, like colonial Australia, ruled by a totalitarian despot. The second group devised a corporate Martian city controlled by a techno-oligarchy inherited from Earth-based companies. Both were hysterically funny visions that left a bad taste in everyone's mouth. After talking through what we didn't like about the two futures we had created, the group concluded that, perhaps, we were going about this all wrong, and that we had presented ourselves with a false dichotomy. Mars did not need cities, but, in Marcela's words, "let's be a nomadic society that values movement over sedentarism and respects its environment as sacred." Neither governments nor private enterprises should lead the way. To celebrate the workshop's conclusion, we drove to a Oaxacan restaurant in the nearby Guerrero neighborhood, drank beer, and ate giant *tlayudas* while we discussed our Martian dreams, as well as our political futures on Earth, as the federal elections that would result in the presidency of Claudia Sheinbaum were to be held the next day.

Mars in Guerrero

A truncated version of Martenochtitlan workshops was held in conjunction with the Intercultural University of the State of Guerrero (UIEG). The idea was to explore the future imaginaries of a more diverse group of participants, in this case, Indigenous students (Nahua, Me'phaa, and Ñuu savi) from rural backgrounds. "It is easy," Amadís stated during the first session with students from Guerrero, "to imagine an astronaut in space eating a hot dog. It is harder to imagine one eating a plate of beans." Sadly, the workshop was cut short because of problems with time and communication: rural areas continue to suffer from intermittent internet connectivity, but the discussions that did take place were highly suggestive, as participants drew from their own cultural and political experiences in the mountains of Guerrero to imagine a future utopia on Mars. The organizers repeated some of the themes of the first two workshops, including a discussion of the importance of myth and the presentation of information about Mars's physical characteristics, but after the first conversations in which the students expressed their interests they decided that the focus of this edition would be the creation of a social contract for life on Mars, a topic that, for reasons of time, had been left out of the first two editions.

Aside from the inevitable references to Hollywood science-fiction movies, much of the discussion revolved around the participants' main concerns: natural resources, social relations, and political organization. They would not build a city, but a small settlement. Students used their own experiences to suggest the kinds of plants that might thrive in the absence of abundant water and to propose the kinds of habitats that might be sustainable on the surface of Mars or underground. They agreed that the social structure should be based on communalism, with rotating responsibilities and a series of rules and sanctions that would preclude the spread of corruption. Many were also interested in the kinds of relations that could be established with any possible alien beings they might encounter, emphasizing the importance of mutual understanding, solidarity, and respect. They explicitly stated their desire to "not make the same choices as the Spanish conquerors," avoiding violence and domination, as well as the individualism they felt characterized contemporary urban society. Another notable omission from this workshop was the notion of mestizaje, which was never mentioned as a way of melding with the environment. Instead, the group elaborated a set of principles for communal existence on Mars: live in harmony with other humans as well as native Martians; establish agreements that emphasize respect and healthy human relations; provide mutual support and solidarity in the event of any crisis or accident; and establish rules that foment the value of environmental care and responsibility.

Moving away from the geographical centrism implicit in the use of the term "Martenochtitlan" and considering the tensions (past and present) between the inhabitants of Guerrero and Mexico City, the encounter was called "Mars in Guerrero," and participants were asked to pick their own name for their Martian settlement. It is telling that the Indigenous students, mostly from the fields of sustainable development and forestry, although a few were studying language and culture, were much less interested in imagining a neo-Mexica culture on Mars than their urban predecessors. The first names they proposed for the settlement did not reference Indigenous culture at all; rather, they spoke more of a desire for the establishment of a utopia in a new world: New Hope, New Mars, and Happiness. Other names did reference contemporary Indigenous place names and notions: Martliaca (a combination of the names Mars and Atliaca, the hometown of several participants), Martli, and Biyu Natse ("Eagle in the process of growing or being born" in Me'phaa).

Some Final Thoughts

Like twentieth-century Mexican sci-fi and other nepantla narratives, Marsarchive.org's activities engage with the baroque, although in a more overtly decolonial register. The workshops oscillate between the "this is serious" frames of planetary science and NewSpace investment in space technology and the "this is play" frames of memes and speculative fiction. While the Mexican space sector attempts to convince the nation that outer-space activities constitute a path to progress and well-being, popular culture reveals the profound ambivalence that hegemonic capitalist modernity inspires in those who do not, or cannot, fully inhabit it, calling attention to capitalism as a "civilizational choice" (Echeverría 2013, 23). Baroque memes and fictional narratives may not "inspire a radical political alternative to capitalist modernity," but, through their theatricality, "on the deep plane of cultural life," they may "manifest the incongruence of this modernity, the possibility and the urgency of an alternative modernity" (36).

Through transgressive transhabitational processes, Marsarchive.org continues to generate spaces of imagination and narration, negotiating the production of visual and textual objects that expand what can be considered space milieux through the establishment of a speculative Mexican presence in outer space. This presence depends on the activation of a combination of what Diana Taylor has termed the "archive" and the "repertoire." Historical narratives are part of a Mexican cultural archive, the inscribed register of a succession of historical periods, accessible in the present, separable from their historical context and the agents of their production, and, of course, the conceptual basis for Marsarchive .org. But through the re-elaboration of archives in the workshops, they become elements of repertoire, performatively activated through a series of interactions and transformations that emerge from the interplay of new actors, knowledges, and interpretations. These transferences "allow for an alternative perspective on historical processes of transnational [or in this case 'transplanetary'] contact" (Taylor 2003, 20).

As Donna Haraway (2016, 12) reminds us, "It matters what stories we tell to tell other stories with." Marsarchive.org restages the narratives and iconography of national memory and identity. But it also restages the dominant astrocapitalist discourse of the "conquest of space" represented by Elon Musk's vision of the human inhabitation of Mars and referenced in the narratives constructed by workshop participants in the form of the Martenochcas' encounter with the ruins of SpaceX City or the description of the Martian enslavers as Martian "corporations." The Mexican scenarios, based on the paradigm of "re-founding" rather than "conquering," participate in a subversive act of transfer; instead of a militaristic or capitalist narrative of domination and extraction, Martenochtitlan, Martelolco, and Mars in Guerrero substitute alternative ecological scenarios that emphasize memory, responsibility, humility, auto-transformation, and collaboration.

MARS STATION

Built in 1892, Estación Marte was originally a stop along the Saltillo–Torreon railway route where cargo trains paused to load strontium carbonate extracted from an open-sky mine twenty miles away. The mining company closed in 1999, and now the town is almost abandoned. Mars Station was named after the locale's most distinctive feature, a low mountain called Cerro Marte for its similarities with the geology of the red planet. UFOs are often sighted there, and inhabitants claim that some "foreigners" wanted to buy the mountain in the 1990s, but the ejidatarios who held the land in common refused to sell. Other foreigners came in 2014 claiming to be from NASA and asking to purchase the mountain to study its magnetism. They say compasses don't work there.

"My Battery is Low and It's Getting Dark" is a 2023 series by Carlos Vielma, originally from Saltillo, inspired, of course, by the viral "last words" of NASA's Opportunity rover, whose battery died after a dust storm on Mars in 2019, as well as Ray Bradbury's (1950) story in *The Martian Chronicles* "A Million Year Picnic" (the one in which the protagonists discover that "the Martians are us"), the film *2001: A Space Odyssey*, and the sense of fear and isolation many people felt during the COVID-19 pandemic. The exposition includes filtered photographs, paintings, drawings, video, and a sculptural wind harp whose sounds evoke Mars Station's melancholy desert landscape. The artist mentions

FIGURE 18 Mars Station. Photo by the Author.

Pedro Páramo, Juan Rulfo's (1955) novel of ghosts and desolation in revolutionary Mexico as a reference for his "rural science fiction" (Fernández,
in Marines 2023).[1]

A few years earlier, in 2010, artists Iván Puig and Andrés Padilla had
decided to mark the celebration of the Bicentenary of Mexican Independence and the Centenary of the Revolution with a project called SEFT-1
(*Sonda de Exploración Ferroviaria Tripulada*, or Crewed Railway Exploration Probe), a critical and aesthetic reflection on Mexico's railway
infrastructure and the promise of progress. Capable of moving along
train tracks and highways, the "rover" SEFT-1 started its journey of exploration in Mexico City. After months of travel, during which time the
artists kept scientific field notes and took photographs of their discoveries at different sites marked by the railways' failed promises, SEFT-1
ended its journey in Coahuila, in the town of Mars Station. For Puig and
Padilla, Estación Marte was the ideal terminus for their mission to probe
the ruins of Mexican modernity.[2]

CHAPTER 6

DARK SKIES

And the thing is that there's a more beautiful sky where we live, because we're in the east, toward Puebla. There the sky is much prettier, right? It's darker. We can appreciate the view of the volcanos, which look cool at night and during the daytime.

—IVÁN, STUDENT

When I was little, we were in darkness a lot. I lived where there were milpas [corn fields]. And, well, you would go out, walk down the street, and everything was open fields. You climbed a tree and could see the sky.

—ERNESTO, TEACHER

Seeds of Light and Darkness

I had originally gotten in touch with CITNOVA, the state of Hidalgo's Council of Science and Technology, because I was interested in the proposed creation of the National Laboratory of Space Access (LANAE), a research center and launch site for stratospheric balloons that would be used to test space technology. The LANAE was to be an offshoot of the UNAM's LINX, the space instrumentation laboratory we saw in chapters 2 and 4. For years, the state government had been promoting Hidalgo as an emerging hub of scientific and technological innovation, touting three flagship projects: the LANAE, a research center dedicated to developing artificial intelligence technology and, more spectacularly, a billion-dollar synchrotron particle accelerator. By the time of this writing, only the Nanomaterials, Robotics, and Artificial Intelligence Laboratory had been inaugurated; the more ambitious projects had been canceled or indefinitely postponed because of budget constraints, local conflicts, and changes in political priorities. When I visited CITNOVA's brand-new campus on the outskirts of Pachuca in 2022, I was struck by a giant work of art, *Synchrotron: Seed of Light* by Chihuahuense artist Miguel Valverde, that hung on the wall of CITNOVA's main lobby. The image is a mandala-shaped atom, radiating all the colors of the visible light spectrum, with a many-petaled "flower of life" in the middle, set against a muted background painted with rabbits, fish, birds, and stars, representing the elements of earth, water, and wind, and painted in the style of traditional Hñahñu iconography. The synchrotron itself will not materialize anytime soon, but the idea of a cutting-edge particle accelerator continues to motivate dreams of a "bright" technological future.

In a conversation with Leonel about the postponement of Hidalgo's megaprojects, the then-director of scientific and technological development at CITNOVA mentioned offhand that they were also collaborating on a community project to protect the night sky. And so, instead of returning to Hidalgo to learn more about projects that would launch technology into space or create light a thousand times brighter than the sun, I found myself traveling to the ejido of San Sebastián, a small community that was asking politicians to turn off the lights.

That visit to San Sebastián would have a profound impact on the way I had been thinking about the territorialities and temporalities of outerspace milieux in Mexico, inviting me to expand my focus beyond the social actors involved in the formal space sector. In this chapter, as a result of my interactions with the "dark-sky community" in Hidalgo, I discuss the interplay between darkness and light that emerges in Mexican engagements with outer space. I show how the connected processes of electrification, astronomical observation, ecological conservation, and community development have been and continue to be shaped by urbanization, immigration, political changes and colonial histories. Dark skies, I argue, represent a multiscalar intersection of technoscientific practice, environmental justice, and community worldmaking.

Enlightenment

In 1615, the Franciscan missionary Juan de Torquemada wrote that the rulers of Tenochtitlan had established a system for illuminating the Mexica capital city using "large braziers of fire along great stretches, and while some slept, others kept watch, so that day and night there would always be someone keeping abreast of what was going on in the city" (in Briseño 2017, 49). This system seems to have been lost after the conquest, as only a few scattered lanterns or burning bundles of pine resin would illuminate the darkness of Mexico City's nights during the first centuries of Spanish rule (50). By the eighteenth century, however, the city government determined that each inhabitant with economic means would be required to place lights outside their house or place of business, according to the time of the month, as moonlight would theoretically complement sources of artificial illumination. These regulations were an attempt to "surveil and regulate nightlife, both for diverse accepted public activities, and for the many other activities that the underworld carried out, taking advantage of invisibility. In this scenario, whoever had light also had power" (52).

Toward the end of the colonial period, the viceroy de Revillagigedo (whom we last saw in chapter 1, overseeing the drainage works in the main square that resulted in the unearthing of the Piedra del Sol) ordered that 1,128 glass lamps be distributed throughout the city's streets and that a significant number of *serenos*, or night watchmen, be employed to keep them burning all night (Briseño 2017, 52).[1] These lights were not merely meant to make nocturnal urban transit more comfortable, but to thwart those inhabitants who might take advantage of the night to commit crimes or indulge in vices; that is, they were used to surveil the poor and Indigenous inhabitants of the city. Revillagigedo believed that more light was needed in the poorer urban areas so that the lower classes, who could not control themselves, "would be [morally] transformed if . . . visible" to the upper classes and the police (Montaño 2021, 53).

During the first century of Mexico's independence, public lighting remained a priority in the new nation's capital city, although "the eternal bankruptcy the country experienced put the brakes on the construction of infrastructure and affected the services the government was able to offer" (Briseño 2017, 54). The system of public lighting also highlighted socioeconomic differences: the center of the city, inhabited by its wealthier residents, was well illuminated while the poorer neighborhoods were left in the dark. Security forces in the form of serenos and policemen were also concentrated in affluent areas.

But between the evening and morning bells that rang out from the Metropolitan Cathedral, illuminated by lamplight or moonlight, city dwellers carried out their nocturnal activities, licit or illicit. At night, although most people were safe inside their homes, public spaces might be occupied by "policemen and *serenos*, midwives, doctors, the wounded, the ill, the moribund and priests; beggars, vagabonds, the wretched and thieves; gamblers, drunks and prostitutes, musicians, actors and singers, insomniacs, the melancholy, writers and poets; the pious and the unfaithful," lovers, conspirators, and all those looking for "the freedom to do as they pleased in complicity with the darkness" (Briseño 2017, 73).

However, most people were safely at home after the official curfew, where grandparents would tell their grandchildren stories about the dangers of being out after dark. These narratives might include

> . . . terrorific traditions, like the ones about Don Juan Manuel, la Llorona, the Mulata of Córdova, and the Coach on fire, or criminal acts, like the murders of Dongo at the end of the 18th century, or scary stories like the Green Mantle of Venice. . . . Or the devil with a tail and horns that emits sparks from his mouth when he speaks. . . . Or a cadaver that walks about on rooftops, threatening the living . . . or *nahuales* that suck children's blood. (In Briseño 2017, 142)

Before they went to sleep, children would kneel beside their beds underneath a cross or virgin affixed to the wall and ask their guardian angels to protect them from all nocturnal dangers, keeping them alive until sunrise (167).

Eventually, oil and then gas lamps replaced candlelight and pine resin (although the candle industry continued to be vibrant, thanks in part to the importance of candles in wakes and other religious ceremonies). But for many, the introduction of electric lighting during the long techno-optimist Porfiriato of the late nineteenth and early twentieth century would truly mark the beginning of Mexican modernity. As Lillian Briseño (2017, 17) writes in her study of electrification during this period, "before Porfirio: war, debacle, insecurity, uncertainty, bankruptcy, political crises, invasions, the loss of territory, darkness; with him, peace, order, progress and light . . ." Electricity would become a fundamental aspect of the project of nineteenth-century modernity, in the same way that satellites would occupy a central place in the discourse of modernity in the middle of the twentieth century.

In 1879, Mexico's first electric plant was built in León, Guanajuato, to power lights and electric looms at the La Americana textile factory. A U.S. mining empresario built Mexico's first hydroelectric plant to serve the silver mines of Batopilas, Chihuahua, ten years later. For the next few years, the mining and textile industries would continue to be the major promotors and main beneficiaries of the new technology, but electricity as a public service would eventually follow. The first experiments with electric lighting in Mexico City took place in 1881, when the Carbon Hydrogen Gas Company installed a series of forty electric arc lights along one of the capital's main thoroughfares, although they were only used on Sundays and holidays (Briseño 2017, 55). However, the installation had mixed results. As one journalist wrote, the blinding arc lights cast such long shadows that the scene looked as if "a world of black giants commanded another world of pygmies illuminated by pale, azure reflections emanated by the moon of a drawn-out cemetery" (Montaño 2021, 36). For some time, even after the introduction of less-blinding illumination technology, electric lights co-habited with other forms of lighting in Mexico City's public spaces; as one member of the city council wrote, walking from downtown to the outskirts of the city was like "flipping through the city's streetlight history" (31). As had been the case for earlier technologies of illumination, "the kind of lighting found in a neighborhood was a clear measure of the class of its inhabitants" (32).

At first, not everyone was enamored of the new technology. Briseño (2008, 24) reproduces a letter published in the periodical *El Imparcial* in 1902 whose author complained that an electrical transformer in front of his business was being used as a toilet and required constant cleaning. And the omnipresence of electrical lights provoked some citizens to lament that "no one can live without sleeping, and the electric lights are killing us" (28). Yet, although many worried about the dangers of electricity, others believed that electricity could cure anything from cancer to hysteria to drug addiction (30). Electricity also implied other dangers, as evidenced in the reporting of the number of injuries and fatalities that resulted from encounters between pedestrians and the city's new electric streetcars (Montaño 2021, 116). Eventually, electricity was normalized, and the blame for streetcar accidents shifted from uncaring tram drivers to ignorant pedestrians, often characterized as "rural bumpkins" who did not know how to navigate the streets of modern Mexico City (138).

The capital's expanding citizenry would come to demand more and more public illumination, convinced that more light would mean greater safety (Briseño 2008, 52). As Diana Montaño (2021, 52) argues, by the beginning of the twentieth century, public electric lighting became synonymous with hygiene, public morality, public order, and public safety. Eventually, electricity would affect all areas of urban life in Mexico: transportation, entertainment, domestic chores, communication, hygiene, and domestic celebrations. The legal system also had to adapt to the new technology, as citizens found ingenious ways to divert electricity from power lines. Lawyers even engaged in metaphysical debates about the nature of electricity, as its characterization either as an intangible force or a material substance would affect the application of property laws (Briseño 2008, 50).

Electricity affected the nation's cultural life, as well. After the triumph of the Revolution, artists and intellectuals who under Porfirio Díaz had been at best ambivalent about industrial technology were eventually won over by its promise as a motor of development and social harmony. Members of the postrevolutionary artistic vanguard known as *estridentismo*, for example, became "foot soldiers of technology" (Gallo 2005, 1), incorporating camaras, typewriters, radios, and other machines as both tools and themes in their works. And, as we have seen, Education Minister José Vasconcelos promulgated the notion of a "cosmic race," which imagined "a future society in which machines and technological artifacts would lead Latin America to triumph over the United States" (5), combining technological improvements with a racial eugenics in which the light of progress was equated with a genetic "lightening" of the dark-skinned population.

The incursion of electricity into the practices of daily life contributed to what Briseño (2008, 26) calls "a culture of light" for most Mexicans. But, as Montaño argues, electrification also implied the juxtaposition of contradictory political, social, and economic forces. The Canadian company Mexlight, which held a virtual monopoly on electrical services, clashed with the Federal Power Commission (CFE), which had been established in 1933, and with members of the Union of Mexican Electricians (SME), who came to be considered anti-imperialist "soldiers of light. . . . Symbolic defender[s] of the electric power consumers and the Mexican people at large" (Montaño 2021, 252). The electrical industry was finally nationalized in 1960, after years of struggle, under the banner "Electricity is Ours!" (258).

But conflicts over light did not only occur between nations. Electrification also reflected and contributed to geographic inequalities within Mexico, as authorities attempted to satisfy the growing demand for electricity in the capital by harnessing the natural resources of other regions whose populations did not, however, enjoy the benefits of the power generated in their territories. Throughout the twentieth century, many communities were forcibly removed from their lands to accommodate the new infrastructure. During the administration of President Carlos Salinas de Gotari (1988–1994), characterized by privatization and neoliberal economic policies, tensions increased between local communities and the CFE, which had attempted to address the need for energy sources through mixed public-private financing, often in collaboration with international institutions like the World Bank (Robinson 2006). Community movements protesting the environmental and social impacts of megaprojects such as hydroelectric dams continue to arise in diverse regions of Mexico, often resulting in new constructions of identity and forms of inhabiting territories (Barabás and Bartolomé 1992; Ibarra 2012; Gómez-Barris 2017), while their adversaries accuse them of being "against progress," forces of the dark against the light of modernity.

Nightfall, with or without electric lights, continues to signal a gendered moral landscape, as well. Women who suffer violence after dark are often held responsible for any violation committed against them. According to the president of the National Institute for Women (INMUJERES), 80 percent of women in 2021 did not feel safe walking at night in Mexico City. One governmental response has been the "Safe Paths" program that, along with paving roads and building crosswalks, aims to install more electrical lighting in those areas of the city considered "unsafe."[2] Electrical illumination has an ambivalent force, therefore, tied to the inclusive notion of "security" in terms of both human rights and the exclusive surveillance of stigmatized populations.

But, eventually, an excess of electrical lighting came to be criticized for other reasons. When listening to the interviews my students did with planetarium visitors, I was struck by how many lamented not being able to see many stars in Mexico City's night sky. Whether they had lived in the capital all their lives or migrated from more rural areas, a profound

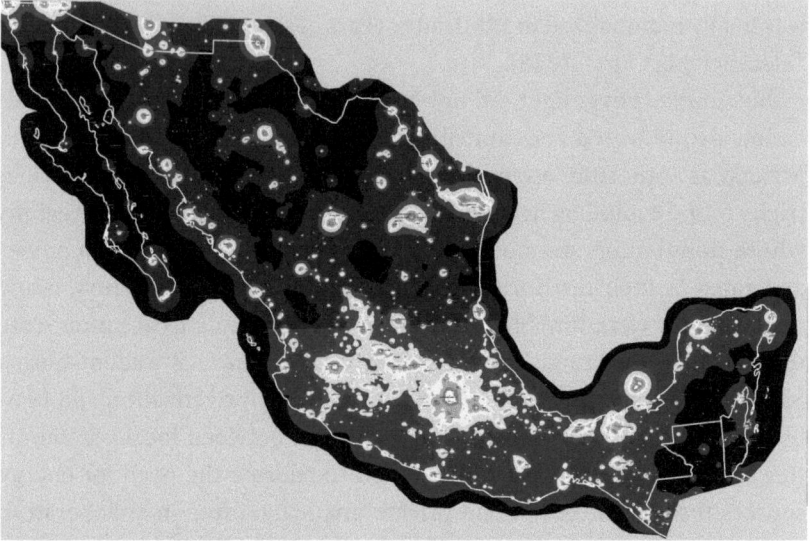

FIGURE 19 Light pollution. Courtesy of sbkmexico.com.

sense of loss seemed to pervade people's testimonies of engagement with the stars. For many in this highly centralized country, life in Mexico City represents economic opportunity. But this comes with a trade-off, as the metropolitan area is the most polluted region in Mexico in terms of air and water quality thanks to its geographical location in a basin surrounded by mountains, as well as to five hundred years of short-sighted urban planning. Increasingly, light pollution is also coming to be seen as an aspect of systemic ecological damage caused by urbanization.

As Mexico City and its surrounding urban zones continue to expand, both territorially and in terms of population density, its "urban stain" on the ground is mirrored by a "radiant stain" in the sky that has been increasingly seen as a problem for humans and other species.[3] Initially, however, only astronomers seemed to care. Indeed, the institutional history of astronomical observation in Mexico has been determined in large part by increasing urbanization and its accompanying luminosity.

Astronomy and the Right to Dark Skies

The first "scientific" astronomical institutions came into being in Mexico in the early nineteenth century. Having won its war against Spain

in 1821, the government of the newly independent country—first the
emperor Agustín de Iturbide and then a succession of presidents of the
Republic—was interested in compiling geographic information about
Mexico, intending to create an integrated map that would take advantage
of regional military and parochial maps and mining plans drawn during
the previous century but update them with "astronomical and trigono-
metrical observations" (Ruiz de Esparza 2003, 56). The Mexican Institute
of Geography and Statistics was formally instituted in 1833. Its members,
assisted by experts from the Mining College and the Military College,
used "practical astronomy" to fix the geographical coordinates of Mexico's major cities and establish the country's borders, including, after the
end of the Mexican-American War in 1848, Mexico's border with the
United States (Bartolucci 2000, 58). An astronomical observatory was
installed in the colonial Palace of Mining in the center of Mexico City in
1840, while the construction of another was begun in Chapultepec Castle. However, the second one was destroyed when the U.S. bombed the
castle in 1847 during its invasion of Mexico. A small observatory built in
San Lázaro in 1856 to support the drawing of more refined atlases was
more successful.

During the next decade, various attempts were made to reanimate the
observatory in Chapultepec, but these were interrupted by the French
invasion of Mexico. The project's leader Miguel Díaz Covarubias, a committed liberal who refused to work for the French, was able to disas-
semble most of the existing instruments and move them out of Mexico
City. This proved fortunate, as the new emperor Maximilian dismantled
what was left of the structure when he decided to use the castle as his
residence. Maximilian later decided to create his own meteorological
and astronomical observatory; however, the agent he sent to buy instruments and equipment in Europe "never returned to Mexico, nor did he
return the money" (Ruiz de Esparza 2003, 59). After the French were
defeated and Maximilian executed, President Benito Juárez and his suc-
cessors supported the idea of establishing a national observatory and
were convinced of the need for astronomical knowledge; for example,
in 1874, a commission of Mexican scientists was sent to Japan to observe the transit of Venus between Earth and the sun. The members of
the commission had great success, publishing their findings in Paris and
gaining international recognition (62). But their plans to finally build a
well-equipped observatory were delayed by political crises in Mexico. A

small observatory had been built on top of the National Palace, but it was useful only for determining the local hour by locating the site's longitude with respect to Greenwich and planning for telegraph routes (60).

The new president, Porfirio Díaz, was advised of the importance of astronomy as an aid to drawing maps that would orient the building of "roads and irrigation works with the object of developing the nation's riches, and simultaneously for the obtention of military data" (Ruiz de Esparza 2003, 61). Finally, in 1878, Díaz inaugurated the National Astronomical Observatory (OAN) in Chapultepec Castle and placed it under the auspices of the Federal Ministry of Development, whose mission was to modernize the Mexican nation, "serve in the formation of the geographic maps," and give "the Mexican intelligences . . . a field of study" (Gallo, in Bartolucci 2000, 60). The observatory's installations would principally support astronomical study for engineering students until it was moved in 1883 to a building in Tacubaya on Mexico City's western border, as, only two years after the first experiments with electrical lighting, the city center had become too brightly illuminated for astronomical observation. In 1903, a new, modern building was built to house the observatory.

This second edition of the OAN was designed by civil engineer and architect Ángel Anguiano, who based the project on the observatories he had seen while visiting Europe to acquire astronomical equipment. In 1929, in a turn away from practical astronomy toward the scientific study of the stars, the observatory came under the auspices of the UNAM. The major project undertaken in Tacubaya was the Carta del Cielo or Sky Map, an international astronomical project on which Mexican scientists collaborated with counterparts at eighteen other observatories located at different latitudes around the world. The goal of the project was the creation of a catalog of stars and the photographic mapping of the visible universe; the first goal was completed in 1970, while the second was never concluded.

The Sky Map telescope and the photographic plates it produced are now considered part of national and international scientific heritage (declared by UNESCO as part of their "Memory of the World" registry in 2014) and are on display at the observatory in Tonantzintla, Puebla, where the OAN was once again relocated in 1951. Today, it shares its location with the governmental National Institute of Astrophysics, Optics, and

Electronics (INAOE), founded after a rift between the UNAM and the famous Mexican astronomer Guillermo Haro, who had been the director of the OAN sites in both Tonantzintla and Tacubaya.[4] Today, a chain-link fence divides the two institutions. From what scientists from both institutions have told me, the OAN got the Carta del Cielo telescope "in the divorce," while the INAOE got the Schmidt camera, also used for astrophotography. Today, these installations are mainly used for teaching purposes, as, again, light contamination has made it difficult to undertake scientific astronomic research.

In 1971, another OAN site was inaugurated in the municipality of San Pedro Mártir in Baja California, one of the country's darkest regions, while the INAOE built a new observatory in Cananea, Sonora.[5] The Ley del Cielo, or "Law of the Sky," a precursor to the federal law defending the right to dark skies, was drafted in Baja California in 2006 to protect astronomical observation at the OAN by regulating public lighting in Ensenada, with another law being passed in 2019 in Mexicali. Attempts to pass similar laws in Tijuana have been unsuccessful (Ávila Castro 2016, 128).

In the years after the passing of the state law, dark-sky advocates broadened the scope of the proposal, arguing that periodic darkness is fundamental not only for scientific advancement, but also for biological health, ecological balance, and economic well-being. They also highlighted the relation between celestial observation and cultural heritage, arguing that being able to see the stars at night is a basic human right.

After a proposal submitted by the Ghanaian and Mexican delegates, the United Nations declared 2015 to be the "International Year of Light and Light-Based Technologies." The next year, in a bid to incorporate the dark side of light-based technologies, UNESCO and the UNAM convened a forum in Mexico, "The Right to Dark Skies," to discuss ways to incorporate the topic of light pollution into national public policy. Astronomers continued to dominate the discussion, but by then dark skies had become an environmental issue and a "human problem" rather than just a scientific one. Eventually, emphasizing the impact of a "rational" use of light on public health and finances, advocates were able to persuade the Mexican Congress to modify the General Law of Ecological Equilibrium and Protection of the Environment, originally drafted in 1988 to address "traditional" environmental problems, to include the management of light pollution, which the law defined as:

The luminous radiance or brilliance in nocturnal environments produced
by the diffusion and reflection of light in gases, aerosols, and suspended
particles in the atmosphere, which alters the natural conditions of luminosity during nocturnal hours and makes astronomical observations of
celestial objects difficult, due to intrusive light, making it necessary to distinguish radiation from celestial sources or objects and the luminescence
of the upper layers of the atmosphere . . .[6]

However, the law is considered by many to have "no teeth," meaning
that it still lacks a body of secondary laws and regulations that would
make it effective. "Politicians have other priorities," one advocate told
me. "Like crime, which tends to be addressed by politicians with more
lights rather than less, and other kinds of environmental contamination."
And, even in Baja California, complications have occurred. The region's
skyglow has been affected by the increasing use of LED lights in public
places and the erection of giant illuminated billboards, as well as the
region's open-pit mining operations, which require large amounts of artificial lighting and generate dust clouds that reflect and amplify light, a
phenomenon that has also affected the INAOE's observatory in Cananea.
Some municipal administrations have adhered to the Ley del Cielo, while
others have not taken it into account when renewing public lighting. Finally, light pollution, like weather, does not obey borders, and the astronomers' night sky in Baja California also depends on lighting policy across
the border in San Diego. While continuing to promote wide-reaching
policy changes, some advocates have focused their attention on smaller-
scale, municipal initiatives.

One such project is underway in Mexico City, where an appreciation
of nature and the night sky has been seen as a remedy for the chaos
of daily urban life. At the "Utopía Libertad," one of a series of "green
utopias" constructed in the southern *alcaldía* of Iztapalapa, visitors can
play soccer or mini golf; feed the goats, chickens, and rabbits at an in-
teractive farm; harvest plants at the urban orchard; swim in the semi-
Olympic-sized pool; enjoy the steam at a temascal; listen to conferences;
check out books from the library; take classes in cooking, yoga, or tai
chi; receive therapy for addictions; or visit the endangered axolotls in the
axolotería. As we saw in chapter 3, they may also experience the wonder
of the cosmos in the Katya Echazarreta Planetarium. After dark, visitors

FIGURE 20 National Observatory, Tonantzintla. Photo by the author.

can observe the night sky through telescopes and learn about the universe, events that draws hundreds of Iztapalapenses to the *utopía* every week. (Appropriately, the Utopía Libertad's symbol is an axolotl peering through a telescope.) The success of the site's stargazing activities, and the 2024 election of the former mayor of Iztapalapa as the head of Mexico City's government, has led to a new project to establish "astronomical utopias" throughout the city. Aside from providing *chilangos* with an escape from the pressures of city life, the utopian project also seeks to counteract the elite access to astronomical knowledge that is a constituent aspect of the construction of national observatories in remote areas.

So far, however, politicians have not taken up DarkSky's banner of "less, better lighting." Marginalized urban populations in Mexico and around the world still suffer more from light pollution and its attendant health risks due to a variety of factors that may include greater surveillance in areas perceived to be "dangerous," the higher population density of economically disadvantaged neighborhoods, and the clustering in these neighborhoods of "undesirable," light-producing commercial and industrial enterprises (Nadybal, Collins, and Grineski 2020, 2). In the case of Mexico City, neighborhoods in the north-central and eastern areas of the capital have the highest levels of light pollution, while southern and western neighborhoods are noticeably darker (Muñoz 2020, 50).[7]

Not surprisingly, dark skies are more easily found far away from the megalopolis. And, unlike resources such as water and energy, the dark-sky landscape cannot be extracted from rural areas and removed to power industry and urban development, although the dark sky as a resource can be commodified.

Peña del Aire

November 2023. I, along with Marcela of Marsarchive.org (whom we met in previous chapters) and astronomer colleagues from my university, gave a Martenochtitlan-inspired workshop called "Mars in Hidalgo" at CITNOVA. We then drove about an hour and a half from Pachuca to the municipality of Huasca de Ocampo, up a narrow, winding highway through piney woods, past two museums, one dedicated to mining and the other to the "*duendes*" or European-style gnomes said to haunt the region since the arrival of English mining companies in the nineteenth century. Another half hour on the road took us past the famous Basalt Prisms admired by Alexander von Humboldt in 1803, the ex-hacienda of the Conde de Regla, and wildflower-studded meadows before we came to the spiny scrub and maguey plants that mark the entrance to Peña del Aire. The site, a wide strip of land bordering the Canyon of Metztitlán, was named after a singular rock formation called variously "the floating stone" or "the Toltec," that broke away from the cliff sometime during the Pleistocene. During the day, the views were breathtaking, and visitors were lined up to ride the zip line or be pushed on the swing that hung over the cliff's edge, perhaps an alternative to the parabolic flights that allow people to briefly experience that exhilaration of weightlessness, defying gravity's pull.

As we ate quesadillas stuffed with *huitlacoche*, mushrooms, chorizo, chicharrón, and cheese, washed down with artisanal pulque, some of the community members told us about strange things they had seen at night. Marcela asked about the ubiquitous duendes. Don Felipe said that a visitor had captured one on film by accident once, and that it looked "just like they paint them for the tourists." More often, though, people see witches in the form of fireballs and strange lights that could also be UFOs. One man showed us a video he had taken on his phone of dogs barking at a strange shape in white. "La llorona?" He shrugged. "How else do you explain it?"

The *socios* of Peña del Aire took us on a brief tour of the site. We walked partway down the cliff beside the Peña. The palapa, a circular hut with a thatched palm roof used for meetings and events, was decorated with a mural of the daytime and nighttime landscape, painted by community members, as well as photographs of local plants and wildlife and posters asking tourists to be respectful of the environment. Inside, three telescopes were arrayed by the window. They had been donated to the community by the UNAM and CITNOVA in support of the new astrotourism project. I asked about the telescopes. "Well, they don't seem to work all that well," I was told. "We've taken a few courses, but you can't really see much more than you can just looking up with your eyes. But you should ask Joshua." "Joshua?" I asked. "The kid from the UNAM. He's in charge of the project." Between 2022 and 2024, I returned to Peña with Joshua many times to give talks at events, help with community workshops, attend meetings with local politicians, and install equipment to measure light pollution.

Ecotourism

The ongoing project to obtain a certification for Peña del Aire as a Dark Sky Park is the latest in a series of initiatives that have seen the Comarca Minera transform from, as its name suggests, a mining region to a productive site for ecological and cultural tourism. Part of the eastern Sierra Madre, the Comarca Minera is considered to have historically produced around 6 percent of all the world's silver, a statistic it owes not only to the presence of the ore, but also to the region's abundant natural resources, particularly timber and water (McKee, Dreier, and Noble 1992). Although first the Teotihuacanos and then the Toltecs had extracted minerals like lime, basalt, and obsidian from the region throughout the pre-Hispanic period, the first mines for the extraction of metal ore were established in Hidalgo by the Spanish in the mid-sixteenth century.

The zone had been an encomienda, or parcel of land whose benefits were granted to a Spanish lord, and then a República de Indios, with certain autonomy for Indigenous leaders. In the eighteenth century, the surrounding lands were consolidated into a hacienda owned by Pedro Romero de Terreros, Conde de Regla, one of the world's richest men at the time, who developed the region's mining industry. The Conde de Regla's administration also saw the first miner's strike in the Americas, when unhappy workers violently resisted a series of measures meant to increase the hacendado's income at the expense of the miners' salaries (Cruz-Domínguez 2012, 68). After the conde's death, mining in the region fell into a period of decline. However, Alexander von Humboldt's visit to the region in the first years of the nineteenth century sparked a wave of international investment in Mexican silver mines, including the incursion of English companies that imported new technologies, like the steam engine that allowed for the pumping of water from deep tunnels, accelerating extractive processes and the consumption of natural resources. They also brought with them folkloric gnomes and Cornish pasties, which became an integral part of local gastronomy. The inhabitants of what is now the ejido of San Sebastián, to which Peña del Aire belongs, historically worked as peons on the mining haciendas while simultaneously farming their own much more precarious fields, located in the barranca. Local inhabitants say that, when they regained their original agricultural lands above the canyon, their patron saint "got heavy," *se hizo pesado*, and didn't want to move, and so they continue to celebrate their most important annual festival in the canyon below.

In response to the "Land and Liberty" and "Land belongs to those who work it" battle cries of the Mexican Revolution that ended the Porfiriato, the old hacienda lands were broken up into communal lands or ejidos. This system of commons governance, enshrined in the federal Constitution of 1917, was the first in the world in which a nation-state recognized collective property. It mandated that rights to the land would be assigned to ejidatarios, but that the ejido would remain the property of the nation and could not be divided, sold, or inherited. San Sebastián was formally constituted as an ejido in 1927, and some of its inhabitants still recall being told how their parents or grandparents had to walk to Mexico City to demand their communal land rights. In 1961, ejidatarios successfully petitioned the government to expand their territory, and in 2001 the ejido was once again ratified in the wake of a controversial 1994 constitutional reform that, while maintaining the concept of communal land rights, made ejidos the legal property of individual ejidatarios, allowing them to sell or rent their parcels.

Subsistence agriculture based on the milpa system (a polycultivate parcel planted principally with maize that also includes complementary species such as beans, squash, chilies, greens, and tomatoes) was the ejido's prime economic activity throughout the twentieth century, at times complemented by small-scale livestock farming and other activities, such as catfish production and the cultivation of fruit trees in orchards and later in greenhouses. However, from the middle of the century onwards, the federal government pursued policies that weakened communal land use in favor of increasing privatization and adherence to the "Green Revolution," a series of technologies promoted by the International Monetary Fund (IMF) and the World Bank meant to increase agricultural productivity in the "underdeveloped world," including the planting of cereal crops resistant to plagues and climate fluctuations, the mechanization of agricultural processes, and the use of genetically altered seeds, chemical fertilizers, and insecticides.

The Green Revolution came at a high financial, social, and ecological cost, contributing to the consolidation of large-scale, industrial agriculture and forcing many campesinos in Hidalgo and other states to immigrate, first to urban areas such as Pachuca and Mexico City, and then to the United States. Immigration reached its peak in the late twentieth and early twenty-first century (Cortes, Granados, and Quezada 2020, 434) and was exacerbated by the North American Free Trade Agreement (NAFTA) of 1994, which made agricultural production in Mexico dependent on market forces (Otero 2004, 77). Sr. David Picazo, born in 1925, six years before the introduction of electric lights in the region, remembered that San Sebastián used to produce a lot of food—maize, nuts, avocados, guayabas—but that a great flood occurred in the 1940s that destroyed the ejido's fields, forcing its inhabitants to immigrate. "Today, the land doesn't give what it gave before. For example, I planted my land, and I took away eighty, ninety, even one hundred sacks [of produce]. But now, you can't even get five sacks. It's because of that famous chemical fertilizer that washed away the soil, making it weak" (Ruiz 2001, 33).

A 1999 report on the ecological and economic situation of the municipality of Huasca de Ocampo suggested that tourism could become a viable path for sustainable development, citing the area's historical and cultural mining heritage, as well as its natural resources, particularly the Barranca of Metztitlán's "biodiversity and aesthetic-scenic" value.[8] The introduction of ecotourism in the region, hailed by local inhabitants as a viable alternative to immigration, occurred in 2000 when the federal government declared the Barranca de Metztitlán to be a protected biosphere reserve. The reserve is noteworthy for its diversity of ecosystems, including xerophytic and submountainous scrub, pine and deciduous tropical forests, grasslands, and riparian vegetation. It is inhabited by 62 species of cactus, 60 mammal species, 215 species of birds, 46 species of reptiles, and 17 species of amphibians, the majority of which are endemic to Mexico.[9] The reserve is also home to a wide variety of nocturnal species, including wild cats, owls, snakes, lizards, and at least twenty-two species of bats. In 2006, UNESCO officially included the reserve in the World Network of Biosphere Reserves.

A more recent initiative by a team from the UNAM's Institute of Geology promoted the certification of the Comarca Minera as an International Geopark, again by UNESCO, because of a series of "internationally relevant" elements, including the epithermal system (Ag-Au) of Pachuca–Real del Monte, the basalt columns of Santa María Regla "described and studied by Alexander von Humboldt in 1803," and the industrial mining heritage that includes colonial haciendas as well as pre-Hispanic obsidian mines (Canet et al. 2017, 4). Toba de Tezoantla, a native rock that geologists classify as a "rhyolitic tuff," has recently been declared an official "heritage rock" by the International Union of Geological Sciences.

It is tempting to see these projects to revalue the region's geology in relation to its biological, cultural, and historical heritage as a challenge to what Elizabeth Povinelli (2016, 17) calls "geontopower," the neoliberal insistence that values existence only with regard to its classification as "life" and ignores what she terms "geontology," the being of geological objects.

FIGURE 21 Peña del Aire. Photo by the author.

Yet institutional projects continue to highlight the potential economic benefits of mineral heritage to humans, rather than meditate on the geological ontologies of "heritage rocks" and their potential for alternative forms of futuring and world-building. (That said, the idea that geological stratigraphy constitutes a territorial "memory" is common among my geologist interlocutors, who may also be secret geontologists.)[10]

The team in charge of the project has established four "geo-routes" (the Humboldt route, the Mining-Historical route, the Geo-cultural route, and the Geo-natural route) that organize the thirty-one official "geo-sites" that make up the park. The purpose of these geo-sites is "the geo-conservation and touristic and educational use of geo-heritage," but they also include sites that reference "biota and ecosystems," as well as "historical-cultural heritage" (Canet et al. 2017, 5).[11] Pre-Hispanic petroglyphs in the Barranca of Metztitlán fall into this category. Many of these carvings reference the importance of the Moon in the Hñahñu vision of the cosmos, as we discussed in chapter 4. The present-day campesinos of San Sebastián are descendants of Nahuas who dominated the Hñahñu before the arrival of the Spanish, but they have incorporated this earlier iconography into their conception of cultural heritage.

Huasca de Ocampo, the seat of the municipality to which San Sebastián belongs, is also included in the category of historical-cultural patrimony, as the first town in Mexico to receive the designation of "*Pueblo Mágico*" (or "Magical Village") from the federal Secretary of Tourism in 2001. This layering of denominations, which weaves together elements of biological, geological, historical, and cultural heritage, has formed the basis for the Comarca Minera's status as a desirable destination for short-term, ecologically focused tourism, especially from Mexico City. And the project to add the international designation of Dark Sky Park to the site of Peña del Aire within the ejido of San Sebastián works to further verticalize historical, cultural, and environmental heritage, from the canyon that makes visible the strata of millions of years of geological transformation and the excavated depths of the defunct silver mines to the cosmic Milky Way.

From Ecotourism to Astrotourism

Turning Peña del Aire into Mexico's first Dark Sky Park made sense to Joshua for several reasons, he told me over coffee and chilaquiles in Mexico City. On the one hand, the site is a prime candidate for the denomination because, despite its relative proximity to urban areas, the light pollution generated in nearby cities is "contained" by the Sierra de Pachuca, resulting in a geological pocket of relative darkness that creates an access point to the night sky. On the other hand, the socios of Peña del Aire have a proven history of community organization, the fundamental "human element" of a dark-sky project. It started when they had to organize themselves to get recognition as an ejido during the period of agrarian reform, but the various ecotourism initiatives of the twenty-first century have resulted in new forms of organization and negotiation. Twenty-six people out of the seventy or eighty formally recognized ejidatarios of San Sebastián created a cooperative of socios to develop and run the site of Peña del Aire. Some sell food or crafts in one of the stalls on the grounds, some charge for the use of the bathrooms, some operate extreme tourism activities (the swing and the zip line), some sell firewood, some give guided hiking or horseback tours, some tell stories around the campfire, and some provide security. Other socios rent cabins to those visitors who wish to spend the night in the area but don't want to camp in tents on-site.

The community has also decided to limit the kind of tourism they promote by moving away from what they call *"turismo de micheladas"* (referring to tourists, understood to be mostly from Mexico City, who lounge around drinking beer with lime and *chile*) and prohibiting the entrance of ATVs. A list of rules posted on the palapa asks visitors to "Care for and respect the biodiversity," "Pack out your trash," "Stay out of the gardens and do not cut flowers," "Refrain from throwing stones into the canyon," and "Keep silent, nature has its own noises." This is the other reason why Joshua thinks Peña will work as a dark-sky park: residents have already developed an ecological consciousness and "culture of environmental protection."

For Peña del Aire to be considered for the international denomination, the site must comply with a series of requirements established by Dark-Sky International (formerly the International Dark Sky Association): there must be public access to the park, the Milky Way must be visible on a typical night, and the site must commit to providing regular dark-sky educational activities. Applicants must also show that they have a long-term management plan, supported by local government, that takes into account what DarkSky calls its "Five Principles for Responsible Outdoor Lighting at Night," which ensures that light is 1) used only when needed, 2) targeted to where it is needed and shielded so that it does not spill beyond this area, 3) only as bright as needed, 4) controlled so that it illuminates only necessary moments, and 5) warm-colored to limit shorter wavelengths of light that scatter in the atmosphere.[12]

Joshua estimates that the certification could occur at the end of 2024 or the beginning of 2025, as his team and community members have spent the last few years assiduously following the process recommended by DarkSky. They measure the quality of the night sky by taking photographs as evidence and retrieving data from the Sky Quality Meters (SQMs) hung in several places along the hiking trail, which had been acquired through a small grant from the UNAM's University Space Program (PEU). The SQMs, Joshua told me, are like simple taxi meters that measure the "behavior of the night sky." They have petitioned the municipal and state governments to apply rational lighting schemes in their jurisdiction (with qualified success), and they have organized many outreach activities, like the projection of films, talks, and workshops about astrophotography, meteor showers, the Moon, Mars, and Saturn.

One of the most important outreach events nationally has been the festival called Noche de las Estrellas, Night of the Stars, celebrated in Mexico City and other sites around the country since 2009. One night a year, hundreds of thousands of participants gather to learn about outer space from universities, astronomy clubs, and other organizations. There is space music, space food, and space theater. But the real draw, even in light-polluted Mexico City, is the array of telescopes, set up in the middle of the venue—usually the main campus of the UNAM—by the organizers and amateurs who bring their own equipment. In 2023, Peña del Aire was an official site of Noche de las Estrellas for the first time. Eric, like Joshua a Dark Sky delegate in Mexico, set up an inflatable planetarium, the astronomy club La Nuit brought telescopes, and experts on a variety of scientific topics gave talks to the community and visitors. There was a "legends campfire" and a nocturnal hiking trail.

One way the community distinguishes Peña del Aire as a dark-sky site, materially and symbolically, is through the display of telescopes, and, to a lesser extent, their use. It makes sense: "dark and quiet skies" were first promoted by astronomers who observe the sky through telescopes, after all, and there are hundreds of astronomy clubs around the country that are always hunting dark skies. The Noche de las Estrellas gained fame when its organizers won the Guinness World Record for most people simultaneously looking at the Moon through telescopes in 2009 and then broke their own record again in 2011. The community mural painted around the palapa features a man gazing at Peña's nighttime sky through a telescope, and telescopes feature prominently in promotional materials. Whenever government authorities come to inspect the site, the telescopes donated by CITNOVA and the UNAM are prominently displayed, and astrotourists expect them to be available. However, as I heard on my first visit to Peña in 2022, using the telescopes can be a frustrating experience for the socios. It isn't quite as easy as "aim at a star and look through the lens."

Joshua and other collaborators, like the astronomers from my university, have offered workshops on the telescopes' use. One participant, a long-time amateur astronomer whose father is from the community, has returned to help residents set up and orient the telescopes, but many are still relatively reluctant to get them out when tourists come. "The truth is," one socio told me, "you can see constellations and experience the night sky much better without them." I have also seen the telescopes pointed at the flora and fauna on the other side of the barranca rather than at the constellations.

A Dark-Sky Community

March 2024. Catherine, a fellow anthropologist working with Joshua, asked a group made up of several generations of ejidatarios to draw a map of their community during a workshop she designed to better understand the construction of material and affective relations between ejidatarios and their environment. They started to draw, adjusting the scale as more elements were added. The river running along the canyon floor to the north of Peña marked the bottom edge of the map, and they filled the outlines of the waterway with drawings of fish, tadpoles, and turtles. A winding path led to the temple of San Sebastián, a corral, and some houses. The southern border (at the top) was marked by the Cerro del Tezontle, locations such as Palma Antigua, El Contento, San Juan Hueyapan, and the old hacienda of Santa María Regla. They included the boundaries of other ejidos. The cartographers marked water sources, the school, some of their houses, caves, cacti, eagles, rabbits, uñas de gato, butterflies, pine trees, snakes, and stars. They paid particular attention to the site of Peña, with the palapa, the zip line, and the Sendero Astronómico drawn in detail. "Do we have to include the other polygons?" one participant asked, referring to the fact that the ejido of San Sebastián is divided into several irregularly shaped parcels, not all of which are adjacent. "Well," responded Catherine, "I think it would be a good idea to include the whole community." There were some eye rolls, and one person muttered that if people from the other polygons weren't present, they shouldn't be included in the map. However, the final product was fairly inclusive. Catherine asked them to stick labels to the map with ideas about their collective future. "There won't be *envidias*," expressions of envy, one person wrote. "We should get along," wrote another. "Paved roads," "growth," "quality of life," and "reforestation" were also mentioned.

In our discussions, Joshua often pointed to pre-existing community organization as an important element of the dark-sky certification process, one that might help the project survive changes in government and public policy like the ones that make dark-sky initiatives in Tijuana and Mexicali difficult. Indeed, one of the things that struck me about the Peña del Aire dark-sky project was Joshua's insistence on the importance of a dark-sky community rather than a dark-sky reserve, in contrast to what he saw as the tendency in places like the United States to locate dark-sky sites in national parks, far from human settlements. What makes Peña special, he said, is not just the quality of the sky at night, but the way that people interact with it in their daily lives.

But "community" implies a complicated web of relations and negotiations. Not every ejidatario is a socio of Peña del Aire, and not every socio is technically an ejidatario. The ejido charges a fee at the entrance of the park, but the ejido does not collect money from activities realized in the park, which are divided among the socios, not necessarily equally. There is a president of the association of Peña del Aire, as well as an elected *comisario* of the ejido, both of whom also interact with the president of the municipality of Huasca de Ocampo and state authorities. Different state offices intervene in sometimes inconsistent ways in Peña's activities. And the fact that most socios are members of the same extended family does not imply that they always get along with each other. Money complicates these relationships, as does friction over decision-making and arguments over rights and obligations. Some socios feel that others receive more tips, or that they go off-script too much when talking to tourists. Some ejidatarios have maintained their distance, while others have decided they want to be involved in the site, "but only now that it's successful.

Where were they when we were working hard to launch the project?" Some of these frictions were brought to light during the celebrations of International Dark Sky Week in April 2023.

We had gone on a night hike with local guides, which had taken longer than expected. When we returned to the palapa, the socios that had promised to bring atole and tamales had already left, to the embarrassment of Ana, the comisaria of San Sebastián, and Angélica, the president of Peña. It also turned out that there had been a charge for going on the night walk, although the posters, sponsored by the state Secretary of Tourism (SECTUR), had advertised that all activities would be free. A representative of the Secretary scolded the organizers, threatening to pull the governor's support if there was any more confusion. Some of the socios had insisted on charging, in defiance of the government's instructions, arguing that, since they were the owners of Peña, they could do as they liked. Others felt that the ones who wanted to charge were being "problematic."

Like other outer-space milieux in Mexico, Peña's dark-sky community requires constant negotiation, between socios and other ejidatarios, but also between local, regional, state, national and international levels of governance. These interactions are necessary, although frictive, and local commons management practices do not always integrate seamlessly with international dark-sky regulations. "Community," therefore, is not a pre-existing organizational structure but a project that must be continually created and recreated, in accordance with shifting relations, pressures, and aspirations, in a cosmopolitical process. If a community is a doing together, an imperfect starting point and an unachievable goal (Magazine 2012, 130), then "commons" are those living systems, those milieux, that make the process of community possible. The Peña del Aire commons, that network of human socios and other biological, mineral, and elemental agents, is irreducible to the form of territorial resource governance known as the ejido of San Sebastián. The commons is always in danger. As Anna Tsing (2015, 135) argues, sometimes the best one can hope for in an imperfect world is the emergence of "latent" or "fugitive" commons, understood as those more-than-human, non-utopian but "good-enough" assemblages that produce entanglements that might be mobilized in common cause.

The Uncanny Night

A few meters away from the palapa where community members were arguing about how to deal with the Secretary of Tourism, Noé, one of the community's best storytellers, took charge of the "legends campfire" that usually accompanies events at Peña del Aire (and whose smoke annoys the astronomers). Noé's stories tend to center on witches, although a duende or a UFO occasionally makes an appearance in his narratives. That night, he explained about witches, who are women who make a pact with the Devil. Men may be naguales, he said, people with a supernatural connection to animals, but only women are witches. They drink the blood of the innocent, particularly children and babies, and they appear as balls of fire. And even if it might seem strange, he told his urban audience, there are more witches in the city than in the country. Think about it: "one light gets lost among all the lights of the city. And besides, there are more hospitals, and more babies being born, more innocent victims." He was lucky because he had a dog when he was young that protected him from the malignant forces that inhabited the night, and from the fireballs he'd seen on occasion.

He still sees things, he said. One night, not so long ago, he had already left Peña to head home when he was called back to the site to "attend a legends campfire." He didn't want to return, but he felt obligated. Strangely, when he got to the campfire, the flames were low, almost extinguished. He saw two fireballs in the camping area, "right over there!" When he got to where he had seen them, they had disappeared, but when he turned around, he saw that the campfire was burning brightly. And when he got back to the campfire, again, the flames had almost gone out. "Forget it," he told himself, "I'm going home. I'm not messing with witches." Some people were less reticent, he told us, referring to the case of a man who trapped a witch in the shape of a fireball. You have to make a cross in the dirt where you see the fireball with the tip of a machete and then stick the machete into the ground. Alternatively, you can take off your underwear, turn it inside out, and throw it at the witch. She'll be trapped. The man who caught one ended up forcing her to marry him, threatening to tell the community her secret. "He had her baptized, got her confirmed, her first communion, and he married her." But after making sure she went through all the sacraments, he sent her away. "Strange things happen at night," he said. "We live in a little village, but you still see a lot of strange things." He told another story about a man who came across a woman washing her clothes at eleven o'clock at night. Strange. He asked her whether she wanted company, and she said yes. He tried to kiss her, but when she turned her face to him, he saw that she had the head of a mule. "He was dumbstruck . . . he never went out at night again."[13]

Laura Romero (2016, 123) writes that, for her Nahua interlocutors in the mountains of Puebla, the night is experienced not only as a moment in time, but as a space that is transformed through darkness. This timespace is uncanny, especially in the absence of electric illumination: strange lights, sinister forces, and beings who do not behave as humans are expected to behave can appear. In Peña, many socios have told me that they have "seen things," although some are more susceptible than others, and recounting and finding meaning in inexplicable nocturnal events has been a recurring topic during my visits to Peña. In this context of uncertainty—which acts at all scales, from corporeal to sociopolitical to cosmic—familiarity with the night and, as Romero argues, "being able to act in it is a sign of power and control over the world, based on a fundamental quality: knowing how to see" (123). She explains,

> Nocturnal spaces become uncertain as our perception changes. Boundaries are erased, as are outlines. The shadows of trees, houses, and people are confused. And so, that which light hides becomes obvious in the dark: the things and beings of the world are more than what they appear at first sight. . . . Eyes cannot see more than what is immediately at hand, and for this reason it becomes necessary to learn to see in the midst of the darkness, but also to hear, to smell, and to feel. Because no one can be sure that a chilling wind that causes your flesh to prickle is not the soul of a dead person who refuses to move on. (126)

Walking at Night

April 22, 2023. Leaving the palapa, a guide gave each of us a fluorescent orange wristband and instructed us to leave our cell phones off so that our eyes would get accustomed to the darkness. Miguel would be in the front, Joshua in the middle, and Marcelo would stay behind the group so we wouldn't get lost. We followed Miguel up the trail and along the edge of the cliff. "Don't get too near the edge," he warned. A few minutes later, we headed away from the drop-off, and the lights from the parking lot and food stalls were hidden behind a slight rise in the land. To our right, the Peña that gives its name to the site was barely visible in the dark. The path was rocky and uneven, and mesquite bushes seemed to shoot out *ahuates*, tiny thorns that scratched my hands and stuck in my clothes. The Moon was new, so I had to pay close attention to the feet of the person ahead of me.

We arrived at Orion, the first *paraje*, or waypoint, and gathered around Miguel, who talked about the constellations we can see in the sky. It had been cloudy, although the rainy season wouldn't begin for a couple of months. But the sky had started to clear. Everyone recognized the three stars in Orion's Belt. An astronomer next to me corrected some of the guide's explanations under her breath. As we prepared to move on, Miguel asked, "Do you want to go the short way or the long way?" The long way would be another two hours of walking, but most of us were willing to make the trek. The few who decided to take the shorter way back headed west, cutting across the middle of the oval loop made by the *sendero*.

The second paraje, Polaris, fittingly marked the trail's northernmost point. The air was cold at two thousand six hundred meters above sea level, although the walk had warmed me up, and the sky was finally clear. "Does anyone know which one is the North Star? Without using your app!" A couple of astronomy enthusiasts pointed it out. We followed the trail around to the southwest. Night noises: the croaking of toads, the sound of crickets, the hooting of an owl, squeaking and rustling in the trees. "Bats," said Marcelo. Indeed, nature has its own noises. *Xoconostle*, large cacti, were silhouetted against the dark, and we maneuvered around *biznagas*, nopales, and *magueyes de luna*. The person in front of me spotted an opossum scurrying into the brush. If there were rattlesnakes, foxes, roadrunners, or *cacomixtles* (ring-tailed cats) in the area, they were staying out of sight.

The third paraje was *La Vía Láctea*. We couldn't see it that night, but Joshua talked about the importance of the Milky Way in Mesoamerican cultures, connecting it to the cosmic tree and Quetzalcóatl. We headed back toward the lights and the hot chocolate. Before we reached our destination, someone yelled, "*La migra!*" Border patrol. Laughter from the students and silence from the guides.

Marcelo talked to Joshua in a low voice. Joshua told me later that he was remembering the three nights he spent walking through the desert in Sonora trying to get across the border to the United States. "It seems to be an obligatory memory during these walks," he reflected. "The first time we marked out the *recorrido*, some of the socios got really emotional when those memories came back."

The Sendero Astronómico, or Astronomic Trail, had been the community's idea, a unique nocturnal educational activity that would attract tourists and satisfy DarkSky's requirement for astronomical outreach. The community had already organized day hikes around the Peña and through the Barranca de Metztitlán as part of their ecotourism activities, so Miguel mapped out a potential nocturnal walking trail that would be safe but slightly challenging, marked with three rest stops named after objects in the night sky. They would charge an extra fee for guiding tourists on these night walks. However, the world of the dark represents a challenge to diurnal mercantile logic. Walkers must become accustomed to seeing less, their hearing is heightened, their feet curl around rocks, their fingers spread out ready to warn of obstacles and ahuates. And walking at night, even in a place that is familiar during the day, makes the familiar strange. Nighttime is when *bolas de fuego*, who are really witches, might be seen, and a walker should be wary. The uncanny night can play tricks on your senses, calling forth embodied memories of other nights in other places, as it did with Marcelo and other socios who were drawn back into memories of dangerous desert border crossings at night, when the darkness was both a threat and protection.[14]

Walking at night, says Ellen Jeffrey (2024, 88), implies a different way of encountering the ground, but it also turns out to be a different way of encountering the sky, in a way that would be impossible in urban settings. In fact, for urbanites, walking in darkness becomes a way of making transparent "the alien nature of how we live in cities" (Dunn 2016, 67). Huicholes say that the planets walk, which distinguishes them from the fixed stars, and Rarámuris say that humans must walk on Earth to maintain order in the cosmos. The Sendero Astronómico plays with these scales and modes of attention. Walking at night is necessarily a mindful practice of environmental immersion. When you stop walking, however, the ground no longer requires your attention, so that you can look up into a dark sky in which, on a cloudless night, the universe seems to stretch away into infinity. After a complete absorption in the earthly milieu and the body's intimate and immediate relation to it comes a lifting of the gaze that situates the body in relation to a wider milieu, small in the face of a cosmic expanse.

Community members say that eco- and astrotourism are alternatives to other economic activities, like immigration. "Now our children don't have to leave home." "Rosa" is Miguel's daughter. She and some of the other kids who live in San Sebastián are almost always present when Joshua and Eric organize activities in Peña. Sometimes she makes TikTok videos about her community. On a recent visit, she told Eric that she had been accompanying her father as he walked around the sendero at night to check on the SQMs placed in the trees. "We're making new memories," both she and her father say. Inspecting the equipment, attentive to the "night's behavior," becomes a process of reworlding, reclaiming the nocturnal landscape from uncanny memories, making it familiar again as a site of present inhabitation. The tourists are the ones who move around, while the socios get to stay where they are. The right to a dark sky is also the right not to immigrate, the right to an emplaced future, and a home in the cosmos.

Some Final Thoughts

While the night sky has always had a presence in the lives of the inhabitants of San Sebastián, its formal recognition as a local "resource" extends the earthly conception of collectivity upward and outward, incorporating the night sky into the terrestrial commons. But the night sky goes beyond colonialist and capitalist logics; it demands a recognition of commons as more than collections of resources. In the spirit of Humboldt's utopian attempts to understand the cosmic planetary as an all-inclusive human and non-human, outer-spatial and terrestrial milieu, the Peña del Aire dark-sky initiative also draws attention to the relations between biological and inorganic beings and things, tied together by complex interactions as well as multiple regimes of value—scientific, economic, ecological, affective, and ontological. Perhaps we might extend Povinelli's concept of geontologies to the (nonmineral) stars? Elementontolgies?

Night "falls," we say in English, and in Spanish, *cae la noche*. Or *anochece*, "it is becoming night." Far from the lights of the city, the thickening of the darkness around us makes it feel as if we are falling into outer space, or as if outer space were falling into us. Recognizing darkness as part of outer-space milieux has a role to play in anticolonial futures and worldmaking practices, in reclaiming the night from the forces of the light, claiming the darkness as commons while resisting having darkness imposed from outside, embracing the dilution of the boundaries between persons and worlds that comes with the night, accepting along with many Indigenous communities that the forces of light are unstable, and that the one-eyed sun will eventually disappear. In its dense weaving of relations between the animal, the vegetable, the mineral, the technological, the elemental, and the otherworldly, the nocturnal milieu creates and is created in a *nepantla* spacetime that demands that we open up our senses and learn to see otherwise.

NOTES

Introduction

1. "Nearshoring" refers to the business practice of transferring processes to nearby countries rather than overseas companies as is the case for "offshoring."
2. Throughout the book, I use real names for public figures but first names or pseudonyms for private citizens, according to my interlocutors' wishes.
3. For more on the intercultural aspects of the municipality of Cuauhtémoc, see Pedroza García (2018).
4. Neri Vela, personal communication, August 2023.
5. See Olson (2018, 82) for the application of the concept of milieu to the relations between human life and technology in outer space systems.
6. I will return to this later, but I will note here that Mesoamerica as an archaeological and ethnohistorical term does not include northern Mexico.
7. This quote is commonly attributed to nineteenth-century dictator Porfirio Díaz, whose stranglehold on power was one of the main causes of the Mexican Revolution. "Poor Mexico," he is said to have lamented when reflecting on the inequalities between his country and its northern neighbor, "so far from God, so close to the United States."

Chapter 1

1. Unless otherwise noted, all translations from the Spanish are my own.
2. Mariel Carpio and Rosa Inés Padilla, at the time anthropology graduate students in the Department of Social and Political Sciences of the Universidad Iberoamericana in Mexico City, conducted the interviews in this section in June 2019.
3. Newspapers in California and Mexico had used the term "*la raza Latina*" from the middle of the nineteenth century in reference to the idea of a shared

Hispanic history and culture, and in 1914 Spain and several Latin American countries began to celebrate *"El día de la Raza"* on October 12 (celebrated in the United States as Columbus Day). Vasconcelos promoted the celebration of the Day of the Iberoamerican Race in 1928, although the name was officially changed in 2020 to the Day of the Pluricultural Nation.

4.   Rivera had famously befriended Trotsky during his exile in Mexico, where the revolutionary had engaged in an affair with the artist Frida Kahlo, Rivera's wife. Siqueiros spent several years in the Lucumberri prison in Mexico City, accused of sedition.

5.   Other programs included the companion project "Mexica Archeoastronomy," as well as films developed by NASA and other digital design companies, like *Return to the Moon* about the Artemis project, *Cosmic Adventure, The Secrets of the Sun, At the Edge of Darkness, The Stars of the Pharaohs,* and *Explorer Robots.*

6.   The film is available at https://www.youtube.com/watch?v=BbRCjHJ0ND8, consulted July 11, 2024.

7.   The German naturalist and traveler Alexander von Humboldt is revered by many historians and scientists in Mexico, especially for the sympathetic portrayal of the nation and critique of colonial structures found in his 1811 work *Political Essay on the Kingdom of New Spain.* However, in the wake of Mary Louise Pratt's 1992 analysis of the relationship between travel writing and European expansion (Pratt 2008), several authors have drawn attention to the more problematic aspects of Humboldt's writings on Latin America. See, for example, the essays in Thurner and Cañizares-Esguerra (2023).

8.   The Mayan Long Count only registered dates until the twenty-first century of our era, giving rise to the "Mayan apocalypse" theory, according to which the world would end in 2012.

9.   Another astronomically based count of 819 days was used by some Classic Maya communities (Aldana y Villalobos 2007, 191).

10. The *Chilam Balam* was compiled from the eighteenth through the nineteenth centuries, but it is probably a copy of an early manuscript from the sixteenth century. The *Codex Mexicanus* has been dated to the last decades of the sixteenth century.

11. Olivier (2006, 174) also writes that the Mexicas considered Cortés's compass to be a kind of mirror.

12. Perhaps coincidentally, an astrologer called Botello, who was a member of the Spanish army and had predicted that the night would end in tragedy, was killed during the rout (Olivier 2006, 187).

Interlude: *Nepantla* Space Program

1.   https://www.e-flux.com/announcements/34310/rigo-23-autonomous-inter galactic-space-program/, consulted July 28, 2024. The image described was designed by Mia Rollow and Tomás.

Chapter 2

1. An earlier version of this chapter was published in the *Routledge Handbook of Social Studies of Outer Space* (Johnson 2023).
2. For an example, see "Anuncio ciudad satélite," https://www.youtube.com/watch ?v=BsF9n48eIw8andt=3s, consulted July 8, 2024. The "city outside the city" was soon swallowed up in Mexico City's urban sprawl.
3. Filmmaker Elena Franco produced an award-winning documentary about the astronaut contest in 2018. *Juanita: Beyond Borders* was based on stories Franco had heard from her grandmother, Juanita Hernández Márquez, one of the sixteen women who appeared on the list of 205 semifinalists. In the film, Juanita proudly tells her granddaughter that she was the only one in her subgroup of ten contestants, all younger than her, to pass all the required tests. "I felt really good," she remembers, with tears in her eyes, "I felt like a *chingona*." She wasn't selected, but having "classified" is still her "main consolation" (Franco 2018).
4. Examples of these kits, they explained, were the satellites launched by academics at the UNAM, widely considered to be the first satellites developed in Mexico. (The director of the UNAM's University Space Program referred to these satellite kits as "off the shelf.") UNAMSAT-1 and UNAMSAT-B (confusingly) were launched from Russia in the 1990s. These satellites were designed to study the trajectories of meteorite impacts. UNAMSAT-1 was destroyed in a launchpad explosion, while UNAMSAT-B was successfully placed in orbit, although it failed after one year owing to a problem with its batteries. Several decades would pass until the next successful Mexican satellite launch.
5. The official was Mtro. Eugenio Urrutia Albisua, Director General of AzTech-Sat-1. https://www.youtube.com/watch?v=YbMcHO_xuoQ&t=1271s.
6. https://astria.tacc.utexas.edu/AstriaGraph/, consulted July 28, 2024.
7. In 2019, Baja California was again the site of a space junk incident, as an artifact possibly belonging to Google's "LOON" rural internet project crashed in the municipality of Comondú ("No era un satélite, era un globo aerostático lo que cayó en Comondú," 2019).
8. Tulancingo 1 is the earlier of two large satellite dishes at the ground station; however, the site's engineers told me, it was built better than it needed to be for the purpose of satellite transmission, so they saved money building Tulancingo 2 to lower standards. But that means that conversion was possible for only Tulancingo 1 (Kurtz et al. 2022).

Interlude: *La NASA no es la raza*

1. https://tresartcollective.com/2020-Trapped, consulted July 28, 2024.
2. Ilana Boltvinik and Rodrigo Viñas, personal communication, November 2024.

Chapter 3

1. Interview with Carmen Félix, November 2018. I looked up NASA's requirements for becoming an astronaut before writing this chapter, and, as Carmen

told me, being a U.S. citizen is the first qualification, listed before having a master's degree in a STEM field, professional experience or flight hours, passing the long-duration astronaut physical, and possessing the "soft-skills" of leadership, teamwork, and communications. https://www.nasa.gov/humans-in-space/astronauts/astronaut-requirements/, consulted July 28, 2024.

2.   https://www.space.com/karman-line-where-does-space-begin, consulted August 3, 2024.

3.   The launch site at Kourou, French Guiana, is operated by the French and European space agencies. See Redfield (2000). Only Brazil operates its own launch sites.

4.   Aside from the three astronauts I mention in this chapter, selected because they were represented on the AEM mural, NASA includes several other Mexican-American astronauts on its list of "Hispanic Astronauts," including Ellen Ochoa, who was the first Latina to fly to space, John D. Olivas, and Sidney M. Gutiérrez. See https://www.nasa.gov/wp-content/uploads/2009/07/hispanic_astronauts_fs.pdf?emrc=aaa679, consulted July 31, 2024.

5.   His story was turned into the film *A Million Miles Away*, produced by Amazon (Márquez 2023).

6.   On other days, the crew was awoken by songs by U2, Gene Autry, the Beatles, Rod Stewart, and Louis Armstrong (Fries 2015, 66).

7.   A statue depicting Hernández as an astronaut was erected in his honor in the Park of Great Values in Mexico State in 2016, but it was stolen by unknown thieves six years later; only the statue's bronze shoes were left behind (Mata 2021).

8.   https://x.com/Astro_Jose/status/989610722253459456, consulted August 1, 2024.

9.   http://www.collectspace.com/ubb/Forum18/HTML/001748.html, consulted July 28, 2024.

10.   Conference, August 2, 2023, Iztapalapa, Mexico City. Available at https://www.youtube.com/watch?v=MUf91r3aIlw.

11.   According to its website, the CSA was founded to underline what its members saw as "a shared dream of freedom [that] brings space and crypto together meaningfully." https://investnews.com.br/financas/brasileiro-viaja-ao-espaco-apos-sorteio-em-compra-de-nfts-privilegio-enorme/, consulted August 2, 2024.

12.   https://spaceforhumanity.org/blog/ca-1announcement, consulted August 1, 2023.

13.   Project PoSSUM (Polar Suborbital Science in the Upper Mesosphere) is now known as PoSSUM 13 in homage to the Mercury 13, a privately funded program in which thirteen women underwent the same tests as the male astronauts involved in NASA's Project Mercury. See https://possum13.org, consulted June 11, 2025.

14.   "Making Humans a Multiplanetary Species." YouTube, uploaded by SpaceX, September 27, 2017, www.youtube.com/watch?v=H7Uyfqi_TE8, consulted July 26, 2022.

15. Activities have been underway for several years to build a Tesla gigafactory in Monterrey. However, the project has been delayed several times for political and economic reasons. Recently, Musk declared that he was pausing the project until after the U.S. presidential election of 2024, as his preferred candidate, Donald Trump, had promised to charge steep tariffs on cars produced in Mexico (*El Financiero* 2024). However, thanks to residential sales and recent government contracts with Musk to provide internet to underserved regions in Mexico, Starlink may be hiring (*Forbes Mexico* 2023).

16. Interview in Mexico City, May 31, 2022.

17. Interview with the members of MEX-1, November 2018.

18. The confusion continues: a recent article about a young astrobiologist who was chosen to command an analog Mars mission in Spain refers to Mónica Ortíz Álvarez, whose accomplishments are indeed notable, as an "astronaut from Oaxaca" and reports that her "mission to Mars was a great moment in her career as an astronaut" (Beltrán 2024).

19. Conversation with Oscar Ojeda over Zoom, August 2, 2022.

20. Colombia does not have a federal space agency. Outer-space activities fall under the responsibility of the armed forces, although some space activities are carried out by a civilian organization.

21. Conversation with "Kevin," Mexico City, June 22, 2022.

22. http://spacegeneration.org, consulted July 30, 2024.

23. Interview by Marcela Chao, https://www.youtube.com/watch?v=tkyfPQxKnZo andt=68s. Consulted August 3, 2024.

24. A young woman who is a member of the Mexican section of the SGAC grouped Mexico with Canada and the U.S., saying that she felt sorrier for Central American and Caribbean countries who did not have the resources of their North American colleagues.

25. In 2023, I was invited to attend the South American regional SGAC meeting in Bogotá, Colombia, where I gave a talk on space anthropology and helped coordinate a working group on Latin American space culture. Although no Mexicans were in attendance, the event was an opportunity to explore the common experiences of Latin American young people in the space field. I also heard a lot of talk about the frictions within the SGAC, especially regarding relations between sections from "emerging countries" and the international leadership.

26. The descriptions and quotes are from the field notes I took at the SGAC national meeting in Tonantzintla, Puebla, July 5–6, 2024.

27. In contrast to the SGAC South American regional event that I attended in Bogotá in 2023, in which members of the Colombian military were active participants, military uses of outer space were not promoted in Tonantzintla, and I did not see a single military uniform. It is common knowledge that the Mexican military operates satellites, although it is not always clear what kind of information they gather. But that is considered separate from "the Mexican space

community," historically guided by the principle of the nonweaponization of outer space.

28. Panel "Sí a la reforma espacial," Senate of the Republic, Mexico City, July 17, 2024.

29. For example, https://www.msn.com/es-mx/pol%C3%ADtica/gobierno/continuar %C3%A1n-con-la-demolici%C3%B3n-dice-rodolfo-neri-vela-sobre-la-agencia -espacial-mexicana/ar-AA1yctgK, consulted February 13, 2025.

30. For example, https://www.facebook.com/story.php?story_fbid=103892119827 7988&id=100064803515953, consulted February 13, 2025.

Interlude: Matters of Gravity

1. Like the title in Spanish, *Asuntos de gravedad / La gravedad de los asuntos*, the project's name links gravity as a physical force with gravity as something that is grave or serious. In English, the title also plays with the ideas of "matter" as material and "matter" in the sense of being important.

2. Miguel Alcubierre is famous for having proposed a mathematical model for faster-than-light travel. The fact that he devised "the Alcubierre Drive" as a graduate student in Wales while watching an episode of *Star Trek* has made him an icon for fans of the sci-fi series.

3. Interview, November 5, 2019.

4. http://www.comisionayotzinapa.segob.gob.mx/es/Comision_para_la_Verdad /Informe_Presidencia, consulted April 28, 2023.

5. Interview, January 20, 2022.

Chapter 4

1. Indeed, the Mexican winning submission in 2015 to the contest held by the International Astronomical Union (IAU) allowing different countries to officially "name" stars and exoplanets was the Nahuatl pair "Tonatiuh" for star HD 104985 and "Meztli" for planet HD 104985 b (https://aui.edu/final-results-of -nameexoworlds-public-vote-released/, consulted February 23, 2025).

2. Interview between Rosa Inés Padilla and Joshua (biochemistry student), Mexico City, June 2019.

3. Interview between Rosa Inés Padilla, Mariel Carpio, Mónica (house cleaner), and Italia (elementary school student), Mexico City, June 2019.

4. The memorandum does not specify the reasons, but apparently Paraguay, Ecuador, and Cuba were the only Latin American countries that did not telecast the event.

5. https://history.state.gov/historicaldocuments/frus1917-72PubDipv08/d29, consulted June 12, 2025.

6. https://www.eluniversal.com.mx/mochilazo-en-el-tiempo/asi-se-vivio-en-la -tierra-la-llegada-la-luna/, consulted August 9, 2024.

7. https://www.youtube.com/watch?v=HMtPJTVxMTEandt=33s, consulted August 9, 2024.

8. https://www.youtube.com/watch?v=mMu_ovEYDNo, consulted August 9, 2024.

9. https://www.eluniversal.com.mx/opinion/mochilazo-en-el-tiempo/el-ingenio -publicitario-de-1969-que-inspiro-la-llegada-del-hombre-a-la-luna/, consulted August 9, 2024.

10. A suggestive oral tradition may date from this visit. Some inhabitants of the state of Colima say that Neil Armstrong was born there, in the town of Zapotitlán de Vadillo, where they reportedly called the future astronaut "El Güero." Novelist Manuel Sánchez de la Madrid wrote about the tradition, which he heard from a domestic worker after watching a documentary on the anniversary of the Moon landings (Sánchez de la Madrid 2010). The 2024 film *Un mexicano en la Luna* is based on Sánchez's novel.

11. https://www.unoosa.org/oosa/en/ourwork/spacelaw/treaties/introouterspace treaty.html, consulted August 15, 2024.

12. See Gómez Revuelta (2025) for more on Mexico's "cosmic diplomacy."

13. https://treaties.unoda.org/t/moon, accessed August 15, 2024.

14. Although his work is only known regionally, a mural painted by Hidalguense artist Jesús Becerril also deserves mention here. His unfinished *Mural of the Angels* was painted in 1972 and can still be seen in the sixteenth-century parish church of the Ascension of Mary in Pachuca, Hidalgo. Inspired by visions of the Apocalypse, the painting centered on an image of the Virgin Mary surrounded by joyful angels. Underneath the angels are four nude male figures, representing the Evangelists, and below them, a plaster statue of the resurrected Christ rises toward Heaven. Surrounding the statue, Becerril painted a scientific vision of the cosmos, complete with stars, comets, galaxies, and a rocket heading toward the Moon. The mural was controversial, however, and never completed—not because of the juxtaposition of Christ and a rocket, but because of the naked male bodies. Becerril explained his impulse to represent the Apollo mission in his religious mural as a desire to represent the historicity of salvation: "I have wanted to represent humanity, and [my work] has a human projection; Heaven is part of reality. We shouldn't separate science from God, although some want to separate them. So, on the altar is a lunar module" (Morales 2005, 48).

15. Two of these "dead rocks" are on display in the astronomy gallery of the UNAM's Universum science museum.

16. The authors discuss whether the relations between China and Latin America should be considered a form of South–South cooperation or an expression of asymmetrical neodependency, only with more diverse actors (Frenkel and Blinder 2020, 6).

17. The "Good Neighbor" policy oriented the United States' relations with Latin America under Franklin Roosevelt. The policy was guided by the principle of nonintervention, while "soft power" was used to promote U.S. interests in the region. During the Cold War, the U.S. did not hesitate to intervene in Latin America to counteract the influence of the Soviet Union. In contemporary dis-

course, the *política del buen vecino* continues to be used by politicians as a way of describing relations between the U.S. and Mexico. See, for example https://www.forbes.com.mx/amlo-recomienda-a-sheinbaum-mantener-una-buena-vecindad-con-eu/, consulted August 20, 2024.

18. The title for this section was inspired by the famous dictum, usually attributed to Porfirio Díaz, "Poor, Mexico, so far from God, so close to the United States!"

19. https://www.gob.mx/sct/prensa/mexico-se-adhiere-al-programa-artemisa-de-la-nasa, consulted August 20, 2024. In the *Handbook for New Space Actors*, Ramírez de Arellano y Haro (2017, viii) states that Mexico was the fourteenth-largest aerospace producer in the world in 2017, although she does not specify how that figure was calculated.

20. Interview between Rosa Inés Padilla and Carlos (engineer) Mexico City, June 2019.

21. https://www.facebook.com/permalink.php?story_fbid=419687615494687an did=362410274555755, consulted August 22, 2024.

22. https://unamglobal.unam.mx/global_revista/la-unam-y-su-rol-en-conquista-del-espacio/, consulted August 26, 2024.

23. https://www.astrobotic.com/update-17-for-peregrine-mission-one/, consulted August 23, 2024.

24. This and subsequent quotations from Gustavo Medina Tanco in this section were taken from a conversation we had in February 2024.

25. See Ochoa (2004) for a discussion of the deployment of the trope of failure in the construction of Mexican national identity.

26. https://x.com/icnunam/status/1748168700748800084, consulted August 26, 2024.

27. https://x.com/icnunam/status/1745501355874119841, consulted August 26, 2024.

28. Octavio Paz (1961, 22) famously wrote that "for us, a realist is always a pessimist." However, while it may be tempting for many of my interlocutors to reduce a reluctance to accept success to an expression of a national inferiority complex conditioned by the historical trauma of the conquest, as many authors writing about *mexicanidad* have done (Ramos 1934; Gamio 1916; Paz 1961), I find it more useful to think of this collective skepticism as a register of political and social consciousness as well as a discursive element of a common sociocultural repertory. See Bartra (1992) and Lomnitz-Adler (1992) for critiques of Mexican national character studies.

29. For thousands of years, the Altar Desert has been inhabited by humans, practicing hunting, foraging, and agriculture. The O'odham are generally considered to be the descendants of the Hohokam. After the conquest of Mexico, the Jesuit Eusebio Kino established missions in the region. The colonial period was marked by violent encounters between the O'odham and Spanish ranchers. In the wake of the 1848 Treaty of Guadalupe Hidalgo, the O'odham were forcibly divided into two, with many of the sacred sites remaining in Mexico, although

a greater part of the population resides in Arizona, a situation that has complicated O'odham movement and organization. See Amador Bech (2005).

30. https://terremoto.mx/simulacros, consulted June 12, 2025.

31. The quotes in this section are from Marcela Chao's interview with Helena Lugo,
available on Marsarchive.org's YouTube channel at https://www.youtube.com
/watch?v=VlKDQt2CU80andlist=PLKg4vAuelqWMj2eVSGn7ePrO5X35dQ
7F0andindex=4, consulted August 29, 2024.

32. Another Nahua story tells of a rabbit who sacrificed himself to feed the god
Quetzalcóatl, who rewarded him by placing his reflection in the Moon so
that humans would remember his generosity. A Rarámuri myth tells of two
poor children who lived alone in a hut while the world was still dark. The first
Rarámuris "cured" the two children with crosses soaked in a drink made from
fermented corn, causing them to glow and illuminate the earth, becoming the
Sun and Moon. A Mayan legend tells the story of Itzmaná and Ixchel, who
became the Sun and Moon after dying tragically for the love of each other.

33. See, for example, Albores and Broda (1997).

Chapter 5

1. It was produced after the demise of the iconic golden age of national filmmak-
ing, but before the emergence of the social critique that characterized the "New
Cinema" of the 1970s. For decades, the government had heavily subsidized the
industry, but a combination of rising costs, dependence on Hollywood, short-
sighted organizational policies, and political crises limited innovation (Berg
1992, 37).

2. Indeed, the moral authority of wrestling *técnicos* continues to provide symbolic
power to social movements in Mexico. The character Superbarrio mobilized
the citizenry in the wake of the 1985 Mexico City earthquake, El Ecologista Universal has protested the building of the Laguna Verde nuclear plant in Veracruz,
as well as other extractivist megaprojects, Super Gay leads the fight against
homophobia, and Fray Tormenta (the only "real" wrestler of the bunch) runs a
home for orphans.

3. Other examples of alien monster movies of the era included *Santo vs the Killers
from Another World* (1971), *Blue Demon and the Female Invaders* (1969), *Viruta
and Capulina vs the Female Astronauts* (1964), and *The Ship of Monsters* (1960).

4. See Morton (2003) for more on twentieth-century Mars cartographic processes.

5. According to the naming conventions of the IAU, large craters on Mars are
named after deceased scientists or writers whose work has contributed to the
study or lore of Mars; small craters are named for villages around the world
(Earth) with populations of fewer than one hundred thousand people, large
valles are called by the names for Mars or "star" in various languages, and small
valles are named after rivers. Tellingly, no large valles have Mexican names, as
Mars was relatively unimportant for pre-Hispanic populations, in contrast to
Venus, which probably has more Mexican toponyms than any other body in

the solar system. See https://planetarynames.wr.usgs.gov/Page/Introduction, consulted September 1, 2024.

6.   The quotes in this section are from a Zoom interview with Marcela Chao, Amadís Ross, and Juan Claudio Toledo, December 1, 2020.

7.   https://www.marsarchive.org/site/, consulted July 2, 2025.

8.   The quotes are all from Marsarchive.org, "Martenochtitlan Primera Fundación," 2020, https://www.marsarchive.org/site/wp-content/uploads/2019/11/5fb958 16e930663dca664512_MartenochtitlanTexto.pdf, consulted June 5, 2024.

9.   In Spanish, there is a play on words between "plumed serpent" (*serpiente emplumado*) and "lead serpent" (*serpiente emplomado*).

10. Workshop "Martenochtitlan: Mito, Rito y Sitio," August 2021.

11. The quotes in this section are from Marsarchive.org, "Mito, Rito y Sitio", 2021. https://www.marsarchive.org/site/wp-content/uploads/2019/11/Mito_2021 .pdf, accessed June 13, 2025.

12. I am exceedingly thankful to the anonymous reviewer of this book for pushing me to further explore Olson's use of the term "transhabitation" and to broaden the range of "transitional practices" involved in the project of Martenochtitlan.

13. A later myth claims that the two groups had been divided ever since the mythical migration from Aztlan: the Mexica deity Huitzilopochtli appeared to each, giving jade to the Tlatelolcos (symbolizing commerce) and fire sticks to the Tenochcas (symbolizing political authority).

14. As Alberto Hernández (2018) observes, the nearby markets of La Merced, La Lagunilla, and Tepito continue the tradition of independent (and informal) commerce in Mexico City to this day.

15. Eugenia Allier-Montaño (2015, 129) writes that, in a survey undertaken in 2007, the "Tlatelolco massacre ranks third as the most remembered date in the country's history (mentioned by 36.2% of respondents), coming after the beginning of the war for independence (49% of respondents), and the start of the Mexican Revolution (39.8%)."

Interlude: Mars Station

1.   See https://www.lanao.com.mx/my-battery-is-low, consulted June 13, 2025.

2.   See https://seft1.net/exploraciones/mexico/monitor/, consulted April 5, 2024.

Chapter 6

1.   The *serenos* also had the unfortunate duty of killing the street dogs that constituted "a plague" in the city and also disturbed citizens' rest with their barking (Briseño 2017, 83).

2.   https://obras.cdmx.gob.mx/proyectos/espacio-publico/senderos-camina-libre -camina-segura, consulted June 13, 2025.

3.   In 1810, when Mexico declared its independence from Spain, the capital city had one hundred fifty thousand inhabitants. One hundred years later, at the start of the Mexican Revolution, the city had quadrupled in size. But urban-

ization accelerated during the second half of the twentieth century, and today Mexico City and its surrounds are home to around twenty-two million people.

4. For a more in-depth recounting of the history of the creation of the OAN and the INAOE, see Elena Poniatowska's (2014) biography of Guillermo Haro, her ex-husband.

5. Sociologist of science Jorge Bartolucci (2000, 20) exasperatedly concludes that astronomy's trajectory in Mexico, particularly when compared with that in the United States and Europe, can be considered "a history of misadventures" whose struggle with economic, political, and intellectual conditions was exacerbated by "sporadic, fragmentary, weak and erratic public and private support." However, newer installations, such as the Large Millimeter Telescope, a radio telescope operated by the INAOE in collaboration with the University of Massachusetts, and the High-Altitude Water Cherenkov Observatory (HAWC), a gamma-ray observatory funded and operated conjointly by institutions from Mexico and the United States, are evidence of the sophistication of contemporary Mexican astronomy and astrophysics.

6. Ley General del Equilibrio Ecológico y la Protección al Ambiente (LGEPA), Diario Oficial de al Federación [DOF], March 28, 1988 (Mex.). https://www .diputados.gob.mx/LeyesBiblio/pdf/LGEEPA.pdf.

7. See Chanda Prescod-Weinstein's (2021) *The Disordered Cosmos* for a discussion of "darkness" in astrophysics and society and Simone Browne's (2015) *Dark Matters* for an analysis of "blackness" as a site of surveillance as well as resistance to racist governance.

8. Consejo Estatal de Ecología del Gobierno del Estado de Hidalgo (1999), *Ordenamiento ecológico territorial de Huasca de Ocampo*, 72. https://bitacora.sema rnath.gob.mx/documentos/huasca_de_ocampo/OET_Huasca_de_Ocampo .pdf.

9. Consejo Nacional de Áreas Protegidas del Gobierno de la República Mexicana (2003), *Programa de manejo de la Reserva de la Biósfera Barranca de Metztitlán*, 11. http://docencia.uaeh.edu.mx/estudios-pertinencia/docs/hidalgo -municipios/Metztitlan-Programa-De-Manejo-Reserva-Biosfera-Barranca-De -Metztitlan.pdf.

10. There are all kinds of "geo-patrimony" in Mexico, including rocks that complicate the distinction between "geo" and "astro," and thus between Earth and cosmos. The meteorite Chicxulub that slammed into what is now the ocean off the coast of Yucatán and did away with the dinosaurs is one. Bacubirito, the fifth-largest meteorite in the world, is another: a twenty-ton rock composed of iron, nickel, cobalt, and other minerals. Found by campesinos in a small town in Sinaloa in 1863, Bacubirito was moved to the civic center in the state capital in 1959 against the campesinos' wishes, and then again to the Center for Science in 1992. It eventually found a home in the interior of the Museum of Sciences in Culiacán. Today it rests in its own "sanctuary," having been declared "Cultural and Historical Patrimony" of the state of Sinaloa. In 2020, Nahum, Manuel, and

Mariana of the KOSMICA Institute (and the eclipse-viewing trip) curated the exhibition *Verses of the Cosmos*, gathering together a group of Mexican artists who had worked with meteorites to explore these "visitors from the universe" that are "more than just rocks: they are encrypted stories from other times and other spaces," as well as "a testimony to the origin and fragility of life in our planetary home" (https://www.kosmicainstitute.com/public_programmes/verses-of-the-cosmos/, accessed June 13, 2025).

11. Emma Elizabeth Ferry (2013, 10) writes about minerals in terms of "an abrupt materiality, a 'thingness' that makes them particularly apt for a study of the production of value in and through objects."

12. https://darksky.org/what-we-do/advancing-responsible-outdoor-lighting/home/, consulted July 16, 2025.

13. Legends campfire, April 22, 2023, Peña del Aire.

14. In 2023, Peña del Aire was nominated for "Best Tourist Experience" in Hidalgo, competing with a nearby site, the El Alberto Ecotourist Park. This park was recognized for its night hike, intentionally designed so that visitors can experience a precarious border crossing for themselves. Members of the local Hñahñu community play the part of narcotraffickers and Border Patrol agents, while tourists must climb rocky hills through cactus-covered paths to cross the Tula River, a stand-in for the Rio Grande.

REFERENCES

Achim, M. 2017. *From Idols to Antiquity: Forging the National Museum of Mexico*. University of Nebraska Press.

Adalid, G. 1985. *Sistema Morelos satélites*. Secretaría de Comunicaciones y Transportes. https://www.youtube.com/watch?v=EfhXLwHu5Og&t=1263s, accessed June 16, 2025.

Aimi, A. 2009. *La verdadera visión de los vencidos: La conquista de México en las fuentes aztecas*. Publicaciones de la Universidad de Alicante.

Albores, B., and J. Broda, eds. 1997. *Graniceros: Cosmovisión y meteorología indígena de Mesoamérica*. El Colegio Mexiquense/IIH/UNAM.

Aldana y Villalobos, G. 2007. *The Apotheosis of Janaab' Pakal: Science, History, and Religion at Classic Maya Palenque*. University Press of Colorado.

Aldana y Villalobos, G. 2021. *Calculating Brilliance: An Intellectual History of Mayan Astronomy at Chichén Itza*. The University of Arizona Press.

Allier-Montaño, E. 2015. "From Conspiracy to Struggle for Democracy: A Historicization of the Political Memories of the Mexican 1968." In *The Struggle for Memory in Latin America*, edited by E. Allier-Montaño and E. Crenzel, 129–46. Palgrave Macmillan.

Alvarado Tezozómoc, F., J. Romero Galván, and G. Díaz Migoyo. 2021. *Crónica mexicana*. UNAM.

Álvarez, C. L. 2023. *Derecho satelital y del espacio exterior*. UNAM/Universidad Panamericana.

Álvarez, R. 1987. "La estación rastreadora de Guaymas." In *Las actividades espaciales en México: Una revisión crítica*, edited by R. Gall and R. Álvarez, 117–20. Fondo de Cultura Económica.

Álvarez, T. 2020. "The Eighth Continent: An Ethnography of Twenty-First Century Euro-American Plans to Settle the Moon." PhD diss. The New School.

Amador Bech, J. 2005. "De la tradición oral a la escritura. Mitos de origen de los O'odham: Versiones, formas de transmission y transcripción." *Anales de la antropología* 1 (39), 131–65.

Anderson, R., E. Backe, T. Nelms, E. Reddy, and J. Trombley. 2018. "Introduction: Speculative Anthropologies." Theorizing the Contemporary, *Fieldsites*, December 18. https://culanth.org/fieldsights/introduction-speculative-anthropologies.

Anreus, A. 2012. "Los Tres Grandes: Ideologies and Styles." In *Mexican Muralism: A Critical History*, edited by A. Anreus, L. Folgarait, and R. A. Greeley, 37–55. University of California Press.

Anzaldúa, G. E. 1987. *Borderlands / La Frontera: The New Mestiza*. Aunt Lute Books.

Anzaldúa, G. E. 2002. "Now Let Us Shift . . . the Path of Conocimiento . . . Inner Work, Public Acts." In *This Bridge We Call Home: Radical Visions for Transformations*, edited by G. E. Anzaldúa and A. Keating, 540–78. Routledge.

Anzaldúa, G. E. 2009. "Let Us Be the Healing of the Wound: The Coyoxauhqui Imperative—la sombra y el sueño." In *The Gloria Anzaldúa Reader*, edited by A. Keating, 303–18. Duke University Press.

Arreola Santander, M. 2017. "Astronautas del programa Apolo en México." *Hacia el espacio*, June 16, 2017. haciaelespacio.aem.gob.mx/revistadigital/articul.php?interior=523.

Astorga Poblete, D. 2014. "Tlacauhtli, altépetl y tlalli: Conceptos básicos de estructuración del espacio, territorio y tierra en el México pre-colombino." *Revista de historia y geografía* 31, 47–61.

Aveni, A. 2012. *Circling the Square: How the Conquest Altered the Shape of Time in Mesoamerica*. American Philosophical Society.

Ávila Castro, F. 2016. "La Ley del Cielo." In *The Right to Dark Skies / El derecho a los cielos oscuros*, edited by N. Sanz, 123–31. UNESCO, Oficina de México.

Ávila Jiménez, N. 2010. *El arte cósmico de Tamayo*. Editorial Praxis/UNAM.

Balibar, É. 2016. *Citizen Subject: Foundations for a Philosophical Anthropology*. Fordham University Press.

Barabás, A., and M. Bartolomé. 1992. "Antropología y relocalizaciones." *Alteridades*, 4, 5–15.

Barnet-Sanchez, H. 2012. "Radical Mestizaje in Chicano/a Murals." In *Mexican Muralism: A Critical History*, edited by A. Anreus, L. Folgarait, and R. A. Greeley, 243–62. University of California Press.

Barrios, J. L. 2022. *Estética y conatus en el arte contemporáneo en México*. Universidad Iberoamericana.

Bartlett, C., M. Marshall, and A. Marshall. 2012. "Two-Eyed Seeing and Other Lessons Learned Within a Co-Learning Journey of Bringing Together Indigenous and Mainstream Knowledges and Ways of Knowing." *Journal of Environmental Studies and Sciences*, 2 (4), 331–40.

Bartolucci Incico, J. 2000. *La modernización de la ciencia en México: El caso de los astrónomos.* UNAM/Plaza y Valdés.

Bartra, R. 1992. *The Cage of Melancholy: Identity and Metamorphosis in the Mexican Character.* Rutgers University Press.

Battaglia, D., ed. 2005. *E. T. Culture: Anthropology in Outerspaces.* Duke University Press.

Battaglia, D. 2014. "Cosmos as Commons: An Activation of Cosmic Diplomacy." *E-Flux*, 58. https://editor.e-flux-systems.com/files/61180_e-flux-journal-cosmos-as-commons-an-activation-of-cosmic-diplomacy.pdf.

Bawaka Country, A. Mitchell, S. Wright, S. Suchet-Pearson, K. Lloyd, L. Burarrwanga, R. Ganambarr, M. Ganambarr-Stubbs, B. Ganambarr, D. Maymuru, and R. Maymuru. 2020. "Dukarr Lakarama: Listening to Guwak, Talking Back to Space Colonization." *Political Geography*, 81, 102218. https://doi.org/10.1016/j.polgeo.2020.102218.

Beltrán, J. F. 2024. "¿Quién es el astronauta Mexicana que logró comandar la mission a Marte?" *Infobae*, January 3, 2024. https://www.infobae.com/mexico/2024/01/03/quien-es-la-astronauta-mexicana-que-logro-comandar-la-mision-a-marte/.

Bennett, J. 2001. *The Enchantment of Modern Life: Attachments, Crossings, and Ethics.* Princeton University Press.

Berg, C. R. 1992. *Cinema of Solitude: A Critical Study of Mexican Film, 1967–1983.* University of Texas Press.

Biggs Coupal, M. 2011. *Exhibiting Mexicanidad: The National Museum of Anthropology in Mexico City in the Mexican Imaginary.* PhD. diss. The University of Texas–Austin.

Bimm, J. 2014. "Rethinking the Overview Effect." *Quest: The History of Spaceflight Quarterly* 1 (31), 39–47.

Bonfiglioli, C. 2008. "El Yúmari, clave de acceso a la cosmología Rarámuri." *Cuicuilco*, 15 (42), 45–60.

Borrego, J., and B. Mody. 1989. "The Morelos Satellite System in Mexico." *Telecommunications Policy* 3 (13), 265–76.

Boyer, D., and G. Marcus. 2020. "Introduction." In *Collaborative Anthropology Today: A Collection of Exceptions*, edited by D. Boyer and G. Marcus, 1–21. Cornell University Press.

Bradbury, Ray. 1950. *The Martian Chronicles.* Doubleday.

Briseño Senosiain, L. 2008. *Candil de la calle, oscuridad de su casa: La iluminación en la ciudad de México durante el Porfiriato.* ITESM/Instituto Mora/Miguel Ángel Porrúa.

Briseño Senosiain, L. 2017. *La noche develada: La ciudad de México en el siglo XIX.* Universidad de Cantabria.

Broda, J. 1995. "Astronomía moderna e historia de la ciencia." In *Historia de la astronomía en México*, edited by M. Álvarez and M. A. Moreno Corral, 44–71. Fondo de Cultura Económica.

Browne, S. 2015. *Dark Matters*. Duke University Press.

Bureaud, A. 2021. "It's a Beautiful Name for a Satellite: Paradoxical Art Objects Somewhere Between Politics and Poetics." *Leonardo* 1 (54), 79–91.

Burkhart, L. M. 1989. *The Slippery Earth: Nahua-Christian Moral Dialogue in Sixteenth-Century Mexico*. University of Arizona Press.

Cabrera López, P. 2013. "Transcendencia del suplemento 'La cultura en México.'" *Impossibilia*, 45–59. https://digibug.ugr.es/handle/10481/41839.

Calvino, Italo. *Invisible Cities*. 1972. Giulio Einaudi.

Campbell, B. 2012. "An Unauthorized History of Post-Mexican School Muralism." In *Mexican Muralism: A Critical History*, edited by A. Anreus, L. Folgarait, and R. A. Greeley, 263–82. University of California Press.

Canet, C., J. C. Mora-Chaparro, A. Iglesias, M. A. Cruz-Pérez, E. Salgado-Marínez, D. Zamudio-Ángeles, E. Fitz-Díaz, R. G. Martínez-Serrano, A. Gil-Ríos, and J. Poch. 2017. "Cartografía geológica para la gestion del geopatrimonio y la planeación de rutas geoturísticas: Aplicación en el Geoparque Mundial de la UNESCO Comarca Minera, Hidalgo." *Terra Digitalis* 2 (1), 1–7.

Canguilhem, G. 2008. *Knowledge of Life*. Fordham University Press.

Carroll, T., D. Jeevendrampillai, and A. Parkhurst. 2017. "Introduction: Towards a General Theory of Failure." In *The Material Culture of Failure: When Things Do Wrong*, edited by T. Carroll, D. Jeevendrampillai, A. Parkhurst, and J. Shackelford, 1–20. Bloomsbury Academic.

Castañeda, L. 2010. "Beyond Tlatelolco: Design, Media, and Politics at Mexico '68." *Grey Room* 40, 100–26.

Castro Sánchez, A. 2019. "Así se vivió en la Tierra la llegada a la Luna." *El Universal*. July 19, 2019. https://www.eluniversal.com.mx/mochilazo-en-el-tiempo/asi-se-vivio-en-la-tierra-la-llegada-la-luna/.

Celorio, G. 2010. "From the Baroque to the Neobaroque." In *Baroque New Worlds: Representation, Transculturation, Counterconquest*, edited by L. Parkinson Zamora and M. Kaup, 487–507. Duke University Press.

Chakrabarty, D. 2021. *The Climate of History in a Planetary Age*. The University of Chicago Press.

Clark, N., and B. Szerszynski. 2022. "Planetary Multiplicity, Earthly Multitudes: Interscalar Practices for a Volatile Planet." In *Narratives of Scale in the Anthropocene: Imagining Human Responsibility in an Age of Scalar Complexity*, edited by B. Dürbeck and P. Hüpkes, 75–93. Routledge.

Coffey, M. K. 2012a. *How a Revolutionary Art Became Official Culture: Murals, Museums, and the Mexican State*. Duke University Press.

Coffey, M. K. 2012b. "'All Mexico on a Wall': Diego Rivera's Murals at the Ministry of Public Education." In *Mexican Muralism: A Critical History*, edited by A. Anreus, L. Folgarait, and R. A. Greeley, 56–74. University of California Press.

Cortés, L. 2023. "Katya Echazarreta: Para competir con un hombre mediocre debes ser una mujer excepcional." *Milenio*. April 29, 2023. https://www.milenio.com/cultura/laberinto/katya-echazarreta-astronauta-naci-deseo-viajar-espacio.

Cortés Rivera, D., J. Granados Alcantar, and M. Quezada Ramírez. 2020. "La migración internacional en Hidalgo: Nuevas dinámicas y actores." *Economía, sociedad y territorio* 63 (20), 429–56.

Cruz-Domínguez, S. 2012. "Conflicto entre trabajadores y mineros de Real del Monte. Antecedentes, documentos y efectos." *Contribuciones desde Coatepec* 23, 67–93.

Cruz Porchini, D. 2021. "Las exposiciones como propaganta estatal: El caso de México en la HemisFair '68 en San Antonio, Texas." In *Diplomacia cultural en México durante la Guerra Fría: Exposiciones y prácticas artísticas, 1946–1968*, edited by D. Cruz Porchini, C. Garay Molina, and M. Velázquez Torres, 77–93. Secretaría de Relaciones Exteriores.

Cruz Porchini, D., C. Garay Molina, and M. Velázquez Torres. 2021. "Introducción." In *Diplomacia cultural en México durante la Guerra Fría: Exposiciones y prácticas artísticas, 1946–1968*, edited by D. Cruz Porchini, C. Garay Molina, and M. Velázquez Torres, 11–18. Secretaría de Relaciones Exteriores.

Dalton, D. 2018. *Mestizo Modernity: Race, Technology, and the Body in Postrevolutionary Mexico*. University of Florida Press.

De la Cadena, M. 2015. *Earth Beings: Ecologies of Practice Across Andean Worlds*. Duke University Press.

De la Puente, A., A. Constantini, F. Torres-Alzaga, G. Esparza, I. Puig, J. J. Díaz Infante, A. Armas, M. Alcubierre, Nahum, and T. Candiani. 2014. "Matters of Gravity." aledela puenteartist.com/_files/ugd/a91104_8bc2c467e23d4f00bf20cbba0ef44abf.pdf.

Díaz Álvarez, A. 2018. "El andar de los días. La cuenta del tiempo entre los grupos del México central, o el llamado Calendario Azteca." *Revista de la Universidad de México* 3, 84–91.

Díaz Álvarez, A. 2020a. "Dissecting the Sky: Discursive Translations in Mexican Colonial Cosmographies." In *Reshaping the World: Debates on Mesoamerican Cosmologies*, edited by A. Díaz Álvarez, 100–39. University Press of Colorado.

Díaz Álvarez, A. 2020b. "Introduction: Rethinking the Mesoamerican Cosmos." In *Reshaping the World: Debates on Mesoamerican Cosmologies*, edited by A. Díaz Álvarez, 3–27. University Press of Colorado.

Díaz Cruz, R. 2023. *El fulgor de la presencia: Ritual, experiencia, performance.* UAM/Gedisa.

Díaz del Castillo, Bernal. 2012 [1568]. *The True History of the Conquest of New Spain*. Translated by Janet Burke and Ted Humphrey. Hackett Publishing Company.

Dillon, Grace. 2012. "Imagining Indigenous Futurisms." In *Walking the Clouds: An Anthology of Indigenous Science Fiction*, edited by G. Dillon, 1–14. University of Arizona Press.

Dovey, C. 2020. "Pale Blue Dot: The Underbelly of the Overview Effect." *The Monthly* 173, 42–47. https://www.themonthly.com.au/issue/2020/december/1606741200/ceridwen-dovey/pale-blue-dot.

Duarte, C. 2021. "¡Vamos a la Luna!: Adhesión de México a los Acuerdos de Artemisa." *Hacia el espacio*. December 15, 2021. https://haciaelespacio.aem.gob.mx/revistadigital/articul.php?interior=1185.

Dunn, N. 2016. *Dark Matters: A Manifesto for the Nocturnal City*. Zero Books.

Echeverría, B. 2013. *La modernidad de lo barroco*. Ediciones Era.

Echeverría, B. 2019. *Modernity and "whiteness."* Polity.

Echeverría García, J. 2015. "'El Sol es comido': Representaciones, prácticas y simbolismos del eclipse solar entre los antiguos nahuas y otros grupos mesoamericanos." *Revista española de antropología americana* 2 (44), 367–91.

El Financiero. 2021. "México quiere llegar sí o sí al espacio y lo hará con ¿Rusia?" *El Financiero*. September 28, 2021. https://www.elfinanciero.com.mx/nacional/2021/09/28/mexico-quiere-llegar-si-o-si-al-espacio-y-lo-hara-con-rusia/?fbclid=IwAR0-wIcDpmuF64mSqCteq09iLCtNS3JmYpDprLsjzdC9PZRAy7pEIxZ3rhU.

Escobar, Arturo. 2018. *Designs for the Pluriverse: Radical Independence, Autonomy, and the Making of Worlds*. Duke University Press.

Evans, B. 2012. *Tragedy and Triumph in Orbit: The Eighties and Early Nineties*. Springer.

Ferdinand, S., I. Villaescusa-Illán, and E. Peeren. 2019. "Introduction. Other Globes: Past and Peripheral Imaginations of Globalization." In *Other Globes*, edited by S. Ferdinand, I. Villaescusa-Illán, and E. Peeren, 1–40. Palgrave MacMillan.

Fernández, C. 2004. "Carlos de Sigüenza y Góngora: Las letras, la astronomía y el saber criollo." *Diálogos latinoamericanos* 9, 59–78.

Fernández-Armesto, F. 1992. "'Aztec' Auguries and Memories of the Conquest of Mexico." *Renaissance Studies* 3–4 (6), 287–305.

Ferry, E. E. 2013. *Minerals, Collecting, and Value Across the U.S.-Mexico Border*. Indiana University Press.

Folgarait, L. 1987. *So Far from Heaven: David Alfaro Siqueiros' The March of Humanity and Mexican Revolutionary Politics*. Cambridge University Press.

Forbes México. 2023. "Gobierno de México revela dos contratos de internet con Starlink de Elon Musk por 3,331 mdd." *Forbes México*. November 15, 2023. https://forbes.com.mx/gobierno-de-mexico-revela-dos-contratos-de-internet-con-starlink-de-elon-musk-por-3331-mdd/#:~:text=El%20contrato%20consiste%20en%20los,de%20datos%2C%20expuso%20el%20Gobierno.

Franco, E., dir. 2018. *Juanita: Mas allá de las fronteras*. NASA/Cinespace.

Freije, V. 2020. *Citizens of Scandal: Journalism, Secrecy, and the Politics of Reckoning in Mexico*. Duke University Press.

Frenkel, A., and D. Blinder. 2020. "Geopolítica y cooperación espacial: China y América del Sur." *Desafíos* 1 (32), 1–30.

Fries, C. 2015. "Chronology of Wakeup Calls." NASA. https://www.nasa.gov/wp-content/uploads/2023/07/wakeup-calls.pdf.

Fuentes, C. 1969. "El espacio o la novela como tonel de las Danaides." *La cultura en México*, August 1969, 8.

Fujigaki Lares, A. 2020. "Caminos rarámuri para sostener o acabar el mundo. Teoría etnográfica, cambio climático y Antropoceno." *Mana*, 26 (1). https://doi.org/10.1590/1678-49442020v26n1a202.

Galindo Trejo, J. 2009. "La astronomía prehispánico como expresión de las nociones de espacio y tiempo en Mesoamérica." *Ciencias* 95, 67–71.

Galinier, J. 1990. *La mitad del mundo: Cuerpo y cosmos en los rituales otomíes.* UNAM/CEMCA/INI.

Gall, R., and R. Álvarez. 1987. "La Comisión Nacional del Espacio Exterior en México, evaluación de sus actividades." In *Las actividades espaciales en México: Una revisión crítica,* edited by R. Gall and R. Alvárez, 108–16. Fondo de Cultura Económica.

Gallo, R. 2005. *Mexican Modernity: The Avant-Garde and the Technological Revolution.* MIT.

Gallo, R. 2020. "Las artes de Tlatelolco." *Confabulario / El Universal.* September 5, 2020. https://confabulario.eluniversal.com.mx/tlatelolco-arte-contemporaneo/.

Gamio, M. 1916. *Forjando patria: Pronacionalismo.* Porrúa.

García Icazbalceta, J. 1891. *Historia de los Mexicanos por sus pinturas.* Díaz de León.

García Ramírez, F. 2021. "Coatlicue, diosa o demonio." *Letras libres.* July 1, 2021. https://letraslibres.com/revista/coatlicue-diosa-o-demonio/.

Garciandía, L. 2017. "The Interplanetary Human." In *Ulises I: An Art Mission to Space by the Colectivo Espacial Mexicano,* edited by J. J. Díaz Infante et al., 31. CONCITEP / Arizona State University.

Gärdebo, J., A. Marzecova, and S. G. Knowles. 2017. "The Orbital Technosphere: The Provision of Meaning and Matter by Satellites." *The Anthropocene Review,* 4 (1), 44–52.

Genauer, E. 1975. *Rufino Tamayo.* Henry N. Abrams.

Godínez-Ortega, L., J. V. Cuatlán-Cortés, J. M. López-Bautista, and B. I. van Tussenbroek. 2021. "A Natural History of Floating *Sargassum* Species (Sargasso) from Mexico." In *Natural History and Ecology of Mexico and Central America,* edited by L. Hufnagel. IntechOpen. https:doi.org/10.5772/intechopen.97230.

Gómez-Barris, M. 2017. *The Extractive Zone: Social Ecologies and Decolonial Perspectives.* Duke University Press.

Gómez Revuelta, G. M. 2025. "Mexican Astropolitics and Scientific Diplomacy, 1958–1977." *Environment and Planning D: Society and Space* 43 (2), 358–79.

González de Bustamante, C. 2012. *'Muy Buenas Noches': Mexico, Television, and the Cold War.* University of Nebraska Press.

González Jácome, A. 2003. "Agricultura y especialistas en ideología agrícola: Tlaxcala, México." In *Graniceros: Cosmovisión y meteorología Indígenas de Mesoamérica,* edited by J. Broda, 467–502. UNAM.

Graham, S. 2016. *Vertical: The City from Satellites to Bunkers.* Verso.

Gruzinski, S. 2010. *Las cuatro partes del mundo: historia de una mundialización.* Fondo de Cultura Económica.

Gruzinski, S. 2012. *La ciudad de México: Una historia.* Fondo de Cultura Económica.

Haraway, D. 2016. *Staying with the Trouble: Making Kin in the Chthulucene.* Duke University Press.

Harman, G. 2010. "Technology, Objects and Things in Heidegger." *Cambridge Journal of Economics* 34, 17–25.

Hecht, G. 2018. "Interscalar Vehicles for an African Anthropocene: On Waste, Temporality, and Violence." *Cultural Anthropology*, 33 (1), 109–41.

Helmreich, S. 2006. "The Signature of Life: Designing the Astrobiological Imagination." *Grey Room* 23, 66–95.

Helmreich, S. 2017. "Foreword: A Wrinkle in Space." *Environmental Humanities* 9 (2), 300–8.

Hernández, J. 2012. *Reaching for the Stars: The Inspiring Story of a Migrant Farmworker Turned Astronaut.* Center Street.

Hernández Hernández, A. 2018. "Tepito, capitalismo a la brava. La tenue frontera entre la legalidad y la ilegalidad." *Alteridades* 55 (28), 99–111.

Howel, E. 2020. "63 Years After Sputnik, Satellites Are Now Woven into the Fabric of Daily Life." *Space.com*, September 29, 2020. https://www.space.com/satellite-technology-daily-life-world-space-week-2020.

Ibarra García, M. V. 2012. "Espacio: Elemento central en los movimientos sociales por megaproyectos." *Desacatos*, 39, 141–58.

Jamasmie, C. 2016. "Infographic: The Facts and Figures That Make Space Mining Real." *Mining.com*, October 6, 2016. https://www.mining.com/infographic-the-facts-and-figures-that-make-space-mining-real/.

Janack, M. 2002. "Dilemmas of Objectivity." *Social Epistemology* 16 (3), 267–81.

Jasanoff, S. 2015. "Future Imperfect: Science, Technology, and the Imaginations of Modernity." In *Dreamscapes of Modernity: Sociotechnical Imaginaries and the Fabrication of Power*, edited by S. Jasanoff and S.-H. Kim, 1–33. University of Chicago Press.

Jeffrery, E. 2024. "Nightfalling: Dancing in the Dark as Artistic Practice." In *Dark Skies: Places, Practices, Communities*, edited by N. Dunn and T. Edensor, 85–93. Routledge.

Johnson, A. W. 2020. "A Mexican Conquest of Space? Cosmopolitanism, Cosmopolitics, and Cosmopoetics in the Mexican Space Industry." *Review of International American Studies* 13 (2), 123–44.

Johnson, A. W. 2023. "Mexico Dreams of Satellites." In *The Routledge Handbook of Social Studies of Outer Space*, edited by J. F. Salazar and A. Gorman, 339–50. Routledge, Taylor and Francis Group.

Juárez, A. 2010. "Siqueiros: La transición de sus paisajes o notas acerca del paisaje en su obra." In *Siqueiros paisajista*, edited by E. Acevedo, M. Fernández, and C. Fulton, 25–33. Editorial R. M.

Karttunen, F., and J. Lockhart. 1976. *Nahuatl in the Middle Years: Language Contact Phenomena in Texts of the Colonial Period.* University of California Press.

Kessler, E. A. 2012. *Picturing the Cosmos: Hubble Space Telescope Images and the Astronomical Sublime.* University of Minnesota Press.

King, R. 2000. "Border Crossings in the Mexican American War." *Bilingual Review / La Revista Bilingüe*, 25 (1), 63–85.

Kirchhoff, P. 2000. "Mesoamérica." *Dimensión antropológica* 19, 15–32.

Kohn, E. 2020. "Anthropology as Cosmic Diplomacy: Toward an Ecological Ethics for Times of Environmental Fragmentation." In *Living Earth Community: Multiple Ways of Being and Knowing,* edited by S. Mickey, M. E. Tucker, and J. Grim, 55–66. Open Book Publishers.

Kohut, M. 2008. "Shaping the Space Age: The International Geophysical Year." *Ask Magazine,* 32, 29–30.

Korpershoek, K. 2023. "Accessibility to Space Infrastructures and Outer Space: Anthropological Insights from Europe's Spaceport." *International Journal of the Commons,* 17 (1), 481–91.

Kurtz, S. E., A. C. Taylor, M. E. Jones, D. M. Gale, E. I. Medel, A. Pollack, and C. Liu. 2022. "A Born Again 32-Meter Radio Telescope for Mexico." *Proc. SPIE* 12182, Ground-based and Airborne Telescopes IX, 121822T. https://doi.org/10.1117/12.2630417.

Lagos, A. 2024. "Gustavo Medina Tanco: 'Este es el inicio de Misión Colmena II, que volará a la Luna en 2027.'" *Wired,* January 18, 2024. https://es.wired.com/articulos/gustavo-medina-no-es-el-fin-es-el-inicio-de-mision-colmena-ii-que-volara-a-la-luna-en-2027.

Latour, B. 2004. "Why Has Critique Run Out of Steam? From Matters of Fact to Matters of Concern." *Critical Inquiry,* 30 (2), 225–48.

Latour, B. 2018. *Down to Earth: Politics in the New Climatic Regime.* Polity Press.

Lempert, W. 2014. "Decolonizing Encounters of the Third Kind: Alternative Futuring in Native Science Fiction Film." *Visual Anthropology Review,* 30 (2), 164–76.

León-Portilla, M. 1962. "Nepantla. La palabra clave de la tragedia de un pueblo." *Excélsior.* January 23, 1962.

León-Portilla, M. 1976. *Culturas en peligro.* Fondo de Cultura Económica.

León-Portilla, M. 1995. "La astronomía y cultura en Mesoamérica." In *Historia de la astronomía en México,* edited by M. Álvarez et al., 4–8. SEP/CONACyT/Fondo de Cultura Económica.

León-Portilla, M. 2017. *La filosofía náhuatl estudiada en sus fuentes.* UNAM.

Lepselter, S. 2016. *The Resonance of Unseen Things: Poetics, Power, Captivity, and UFOs in the American Uncanny.* University of Michigan Press.

Levi, H. 2008. *The World of Lucha Libre: Secrets, Revelations, and Mexican National Identity.* Duke University Press.

Lezama Lima, J. 1993 [1957]. *La expresión americana.* Fondo de Cultura Económica.

Litfin, K. 1999. "The Status of the Statistical State: Satellites and the Diffusion of Epistemic Sovereignty." *Global Society* 13 (1), 95–116.

Lomnitz-Adler, C. 1992. *Exits from the Labyrinth: Culture and Ideology in the Mexican National Space.* University California Press.

Lomnitz-Adler, C. 1996. "Ritual, rumor y corrupción en la formación del espacio nacional en México." *Revista mexicana de sociología* 58 (2), 21–51.

López Austin, A. 1973. *Hombre-dios. Religión y política en el mundo náhuatl.* UNAM.

López Austin, A. 1980. *Cuerpo humano e ideología. Las concepciones de los antiguos nahuas.* UNAM.

López Austin, A. 1996. *The Rabbit on the Face of the Moon: Mythology in the Meso-american Tradition*. University of Utah Press.

López Luján, L. 2008. "'El adios y triste queja del gran Calendario Azteca': El incesante peregrinar de la Piedra del Sol." *Arqueología mexicana* 16 (91), 78–83.

López Velarde, L. 2023. "Generación del sistema jurídico espacial mexicano." In *La nueva era espacial: Cooperación y regulación*, edited by M. Arreola Santander, 28–51. Gallardo Ediciones.

Lorente y Fernández, D. 2011. *La razzia cósmica: Una concepción Nahau sobre el clima. Deidades del agua y graniceros en la sierra de Texcoco*. CIESAS/Universidad Iberoamericana.

Lorente y Fernández, D. 2023. "El nacimiento del Sol y de la Luna entre los nahuas de Texcoco y de la Sierra Norte de Puebla (México)." *Revista española de antropología americana* 53 (1), 217–23.

Mack, P. 1990. *Viewing the Earth: The Social Construction of the Landsat Satellite System*. MIT Press.

Maffie, J. 2003. "To Walk in Balance: An Encounter Between Contemporary Western Science and Conquest-Era Nahua Philosophy." In *Science and Other Cultures: Issues in Philosophies of Science and Technology*, edited by R. Figueroa and S. Harding, 70–90. Routledge.

Maffie, J. 2013. *Aztec Philosophy: Understanding a World in Motion*. University Press of Colorado.

Magazine, R. 2012. *The Village Is Like a Wheel: Rethinking Cargos, Family, and Ethnicity in Highland Mexico*. University of Arizona Press.

Manilla, J. 2024. "Los famosos satélites de Tulancingo, primeros en Latinoamérica, en esto se convertirán." *La Silla Rota*, November 10, 2024. https://lasillarota.com/hidalgo/reportajes/2024/10/11/los-famosos-satelites-de-tulancingo-primeros-en-latinoamerica-en-esto-se-convertiran-505342.html.

Marcus, G. 2000. "Introduction." In *Para-sites: A Casebook Against Cynical Reason*, edited by G. Marcus, 1–14. University of Chicago Press.

Marines, M. "Carlos Vielma encuentra marcianos en el desierto de General Cepeda." *Vanguardia*, April 11, 2023. https://vanguardia.com.mx/show/artes/carlos-vielma-encuentra-marcianos-en-el-desierto-de-general-cepeda-YA7174346.

Marquez Abella, A. dir. 2023. *A Million Miles Away*. Amazon Prime Video.

Marsarchive.org. 2023. *Martelolco: Memorias portátiles de un futuro en Marte*. https://www.marsarchive.org/site/wp-content/uploads/2019/11/ZineMartelolcoLecturaWEB.pdf.

Martínez Mendoza, J., and S. Palomares Sánchez. 2004. "La aventura en Cabo Tuna. Una historia de lo que pudo haber sido." *Entrelíneas* 11 (1), 6–7.

Martínez Ramírez, M. I. 2021. "El camino y el caminar: Fuentes históricas de los rarámuri de la Sierra Tarahumara, México." *Revista de antropologia*, 64 (1), e184479.

Mata, M. 2021. "Roban estatua en honor al astronauta mexicano José Hernández, en Zinacantepec." *Milenio*. June 26, 2021. https://www.milenio.com/policia/zinacantepec-roban-estatua-astronauta-jose-hernandez-moreno.

McCaa, R. 2000. "The Peopling of Mexico from Origins to Revolution." In *A Population History of North America*, edited by M. R. Haines and R. H. Steckel, 241–304. Cambridge University Press.

McKee, E. H., J. E. Dreier, and D. C. Noble. 1992. "Early Miocene Hydrothermal Activity at Pachuca-Real del Monte, Mexico: An Example of Space-Time Association of Volcanism and Epithermal Ag-Au Vein Mineralization." *Economic Geology*, 87 (6), 1635–37.

Medina, C. 2003. "La lección arquitectónica de Arnold Schwarzenegger." *Arquine* 23, 68.

Mejuto González, J. 2023. "¿Hay espacio para todos? Etnicidad y acceso a la tecnología espacial." *Ciencias espaciales* 14 (1), 52–59.

Méndez-Fierros, H. 2023. "La frontera inteligente Estados Unidos–México. Representaciones de tecnología y construcción del migrante irregular como amenaza–enemigo." *Estudios fronterizos*, 24. https://doi.org/10.21670/ref.2317128.

Messeri, L. 2016. *Placing Outer Space: An Earthly Ethnography of Other Worlds*. Duke University Press.

Mezzadra, S., and B. Neilson. 2013. *Border As Method, or, The Multiplication of Labor*. Duke University Press.

Mikulska Dabrowska, K. 2008. "El concepto de 'ihuicatl' en la cosmovisión nahua y sus representaciones gráficas en códices." *Revista española de antropología americana* 38, 151–71.

Mikulska Dabrowska, K. 2020. "The Sky, the Night, and the Number Nine: Considerations of the Nahua Vision of the Universe." In *Reshaping the World: Debates on Mesoamerican Cosmologies*, edited by A. Díaz Álvarez, 264–94. University Press of Colorado.

Milbrath, S. 1999. *Star Gods of the Maya: Astronomy in Art, Folklore, and Calendars*. University of Texas Press.

Miranda, F. 2025. "Semarnat y profepa toman evidencia de daño ambiental por basura especial de SpaceX." *Milenio*. June 23, 2025. https://www.milenio.com/estados/semarnat-profepa-toma-evidencia-dano-ambiental-space.

Mitchell, S. T. 2017. *Constellations of Inequality: Space, Race, and Utopia in Brazil*. The University of Chicago Press.

Monsiváis, C. 1969. "Ya no escribas Julio Verne que hoy sólo rifa Arthur C. Clarke." *La cultura en México*, August 1969, 10.

Monsiváis, C. 1986. "El día del derrumbe y las semanas de la comunidad." *Cuadernos políticos* 45, 11–24.

Monsiváis, C. 1988. *Entrada libre. Crónicas de la sociedad que se organiza*. Ediciones Era.

Monsiváis, C. 2009. *Los rituales del caos*. Ediciones Era.

Montaño, D. J. 2021. *Electrifying Mexico: Technology and the Transformation of a Modern City*. University of Texas Press.

Montero García, I. 2022. *La astronomía en México*. iTiO Ediciones.

Morales Damián, M. 2005. "Formas artísticas y creencias religiosas en la parroquia de la Asunción de Pachuca." In *Arcanos hidalguenses: En memoria de Víctor Man-*

uel Ballesteros García, edited by E. Rivas Paniagua and E. Luvián Torres, 43–50. Universidad Autónoma del Estado de Hidalgo.

Morton, O. 2003. *Mapping Mars: Science, Imagination, and the Birth of a World.* Picador.

Morton, O. 2019. *The Moon: A History for the Future.* Public Affairs.

Morton, T. 2013. *Hyperobects: Philosophy and Ecology After the End of the World.* University of Minnesota Press.

Mulato, A. 2017. "El joven astronauta mexicano no va a Marte, va a un simulador en Utah." *El Pais.* February 13, 2017. https://verne.elpais.com/verne/2017/02/13 /mexico/1487015092_693856.html.

Mullan, C. 2018. "fDi's Aerospace Cities of the Future 2018/19—The Winners." *fDi Intelligence.* August 14, 2018. https://www.fdiintelligence.com/content/4dff03d6 -fd51-51d6-b408-b849ef94b126.

Mundy, B. E. 2023. "The 2022 Josephine Waters Bennett Lecture: Mexica Space and Habsburg Time." *Renaissance Quarterly* 76 (2), 365–407.

Muñoz, J. 2020. *Estudio de la luz artificial durante la noche en la Ciudad de México.* Master's thesis. Universidad Nacional Autónoma de México.

Nadybal, S. M., T. W. Collins, and S. E. Grineski. 2020. "Light Pollution Inequities in the Continental United States: A Distributive Environmental Justice Analysis." *Environmental Research*, 189, 109959. https://doi.org/10.1016/j.envres.2020 .109959.

Navarrete Linares, F. 2019. "Aztec Monoliths as Time-Shaping Devices." *Revista de antropología* 62 (3), 744–68. https://doi.org/10.11606/2179-0892.ra.2019.165226.

Navarrete Linares, F. 2021. "La cosmohistoria: Cómo construir la historia de mundos plurales." In *Cosmopolítica y cosmohistoria: Una antisíntesis*, edited by J. Neurath and M. I. Martínez Ramírez, 25–43. SB Editorial.

Neri Vela, R. 1992. *Vuelta al mundo en noventa minutos.* Editorial Atlántida.

Neri Vela, R., dir. 1993. *México llamando Atlantis: Documental sobre el primer y único astronauta mexicano.* https://www.youtube.com/watch?reload=9&v=wkI6 _BgU8Is.

"Neri Vela quiere volver al espacio." *Expansión.* April 18, 2009. https://expansion .mx/actualidad/2009/04/16/el-astronauta-del-siglo-xx#:~:text=No%20s%C3%A9 %20con%20cu%C3%A1ntos%20hispanos%20concurs%C3%B3%20%C3%A9l%2C %20pero,las%20que%20se%20iban%20a%20colocar%20sat%C3%A9lites%20mexi canos.

Neurath, J. 2004. "El doble personaje del planeta Venus en las religiones indígenas del Gran Nayar: Mitología, ritual agrícola y sacrificio." *Journal de la société des américanistes* 90 (1), 93–118.

Neurath, J. 2021. "Cosmopolítica contra biopoder: Vida, poder y autonomía entre los wixaritari." In *Cosmopolítica y cosmohistoria: Una antisíntesis*, edited by M. I. Martínez Ramírez and J. Neurath, 215–41, SB Editores.

Neurath, J. 2023. *Las religiones indígenas de Mesoamérica: Historia, ritos y transformaciones.* SB Editores.

Nielsen, J., and T. Sellner Reunert. 2020. "Colliding Universes: A Reconsideration of the Structure of the Precolumbian Mesoamerican Cosmos." In *Reshaping the World: Debates on Mesoamerican Cosmologies*, edited by A. Díaz Álvarez, 37–72, University Press of Colorado.

Nieto Calleja, R. 2016. "Trabajo en la globalidad hegemónica. Performance laboral en México y Guatemala." *Revista andaluza de antropología*, 11, 16–43.

"No era un satélite, era un globo aerostático lo que cayó en Comondú." 2019. *La Silla Rota*, May 3, 2019. https://lasillarota.com/estados/2019/3/5/no-era-un-satelite -era-un-globo-aerostatico-lo-que-cayo-en-comondu-180992.html.

Nora, P. 1989. "Between Memory and History: Les Lieux de Mémoire." *Representaciones* 26, 7–24.

Ochoa, J. 2004. *The Uses of Failure in Mexican Literature and Identity*. University of Texas Press.

O'Hara, M. 2018. *The History of the Future in Colonial Mexico*. Yale University Press.

Olivier, G. 2006. "Indios y españoles frente a prácticas adivinatorias y presagios durante la conquista de México." *Estudios de cultura náhuatl* 37, 169–92.

Olivier, G. 2019. "Tetzáhuitl: Los presagios de la conquista de México." *Arqueología mexicana* 89, 28–53.

Olson, V. 2018. *Into the Extreme: U.S. Environmental Systems and Politics Beyond Earth*. University of Minnesota Press.

Olson, V. 2023. "Refielding in More-Than-Terran Spaces." *The Routledge Handbook of Social Studies of Outer Space*, edited by Juan Francisco Salazar and Alice Gorman. Routledge.

Olson, V., and L. Messeri. 2015. "Beyond the Anthropocene." *Environment and Society*, 6, 28–47. http://www.jstor.org/stable/26204949.

Otero, G. 2004. *¿Adios al campesinado? Democracia y formación política de las clases en el México rural*. Universidad Autónoma de Zacatecas / Simon Fraser University.

Pang, A., and B. Twiggs. 2011. "Citizen Satellites." *Scientific American* 304 (2), 48–53.

Parks, L. 2005. *Cultures in Orbit*. Duke University Press.

Pastrana Flores, G. 2004. *Historia de la Conquista: Aspectos de la historiografía de tradición náhuatl*. UNAM.

Paz, O. 1961. *The Labyrinth of Solitude*. Grove.

Paz, O. 1969. "Paschimottanasa." *La cultura en México*, August 1969, 9.

Paz, O. 1995. *Los privilegios de la vista II*. Fondo de Cultura Económica.

Pedroza García, R. 2018. "Cuauhtémoc, Chihuahua: ¿La Ciudad de las Tres Cutluras? Ejemplo de una comunidad imaginada en el Norte de México." *Nueva antropología* 31 (89), 24–42.

Pisano, N. 2024. "How Taco Bell Tortillas Made an Impact on Space Food." *Mashed*. January 28, 2024. https://www.mashed.com/1499703/taco-bell-tortillas-impact -enhance-space-food/.

Platzi. 2023. "Entrevistando a una astonauta, ft. Katya Echazarreta." May 16, 2023. https://www.youtube.com/watch?v=n_vWDdqnnfE.

Poniatowska, E. 1975. *Massacre in Mexico*. Viking Press.

Poniatowska, E. 2014. *El universo o nada: biografía del estrellero Guillermo Haro*. Seix Barral.

Povinelli, E. A. 2016. *Geontologies: A Requiem to Late Liberalism*. Duke University Press. https://www.dukeupress.edu/geontologies.

Pratt, M. L. 2008 [1992]. *Imperial Eyes: Travel Writing and Transculturation*. Routledge.

Prescod-Weinstein, C. 2021. *The Disordered Cosmos: A Journey into Dark Matter, Spacetime, and Dreams Deferred*. Bold Type Books.

Questa, A. 2019. "Broken Pillars of the Sky: Masewal Actions and Reflections on Modernity, Spirits, and a Damaged World." In *Indigenous Perceptions of the End of the World: Creating a Cosmopolitics of Change*, edited by R. Bold, 29–50. Palgrave MacMillan.

Quijano, A. 2000. "Coloniality of Power, Eurocentrism, and Latin America." *Nepantla: View from South* 1 (3), 533–80.

Ragot, N. 2016. "Ritos nocturnos y nacimiento del sol entre los aztecas." In *Las cosas de la noche: Una mirada diferente*, edited by A. Becquelin and J. Galinier, 74–86. Centro de estudios mexicanos y centroamericanos.

Ramírez de Arellano y Haro, R. 2017. "Preámbulo." In *Manual para nuevos actores en el espacio*, edited by C. Johnson, viii–ix. Secure World Foundation.

Ramos, S. 1934. *El perfil del hombre y la cultura en México*. Imprenta Mundial.

Redfield, P. 2000. *Space in the Tropics: From Convicts to Rockets in French Guiana*. University of California Press.

Reyes, N. 2017. "Cuando los conquistadores de la Luna llegaron a México." *El Universal*. October 5, 2017. https://www.eluniversal.com.mx/colaboracion/mochilazo-en-el-tiempo/nacion/sociedad/cuando-los-conquistadores-de-la-luna-llegaron/.

Riva Parga, J. R. 2017. "La exploración espacial: una oportunidad para incrementar el poder nacional del estado mexicano." *Revista del Centro de Estudios Superiores Navales* 38 (4), 33–62.

Robinson, Kim Stanley. 1992. *Red Mars*. Spectra.

Robinson, Kim Stanley. 1993. *Green Mars*. Spectra.

Robinson, Kim Stanley. 1996. *Blue Mars*. Spectra.

Robinson, S. S. 2006. "CFE: Cambio a reversa o la "nueva" vieja política de desalojos forzosos." *La jornada ecológica*, March 27, 2006. https://www.jornada.com.mx/2006/03/27/eco-c.html.

Rodríguez, N. 2025. "México en órbita: Sheinbaum anuncia la creación del Programa Espacial Mexicano." *Ecoosfera*, January 31, 2025. https://ecoosfera.com/noticias/mexico-en-orbita-sheinbaum-anuncia-la-creacion-del-programa-espacial-mexicano/.

Romero, L. 2016. "Pueblo diurno, pueblo nocturno: las nociones Nahuas sobre la noche y la oscuridad." In *Las cosas de la noche: Una mirada diferente*, edited by A. M. Becquelin and J. Galinier, 123–28. Centro de Estudios Mexicanos y Centroamericanos.

Rosas Mantecón, A. 2021. "El Museo Nacional de Antropología de México." *Culturas* 14, 129–44.

Rubenstein, M.-J. 2022. *Astrotopia: The Dangerous Religion of the Corporate Space Race.* University of Chicago Press.

Ruíz, M. 2001. *Nuestros abuelos: La historia viva de Huasca.* PACMYC.

Ruíz de Esparza, J. 2003. "Los orígenes del observatorio astronómico nacional." *Ciencias* 69, 54–63.

Rulfo, Juan. 1955. *Pedro Páramo.* Fondo de Cultura Económica.

Sage, D. 2008. "Framing Space: A Popular Geopolitics of American Manifest Destiny in Outer Space." *Geopolitics* 13 (1), 27–53.

Salas Cuesta, M., and J. Talavera González. 2010. "Una visión de la vida y de la muerte en el México prehispánico." *Arqueología mexicana* 17 (102), 18–23.

Salazar, J. F. 2017. "Speculative Fabulation: Researching Worlds to Come in Antarctica." In *Anthropologies and Futures: Researching Emerging and Uncertain Worlds,* edited by J. F. Salazar, S. Pink, A. Irving, and J. Sjöberg, 151–70. Routledge.

Sánchez de la Madrid, M. 2010. *Un mexicano en la luna.* Diario de Colima.

Sanchis Amat, V. 2020. *"Y todo esto pasó con nosotros": Reescrituras del mundo indígena en la recepción literaria de Tlatelolco 1968.* Iberoamericana Vervuert.

Schmelz, I. 2022. *Codigofagia: Cine mexicano y ciencia ficción.* Akal.

Scott-Heron, G. 1970. "Whitey on the Moon." In *Small Talk at 125th and Lenox,* by G. Scott-Heron. Flying Dutchman/RCA.

Secretaría de Comunicaciones y Transportes (SCT) / Gobierno de México. n.d. "Rumbo al espacio. Sistema de satélites Morelos." https://elmirador.sct.gob.mx /cuando-el-futuro-nos-alcanza/rumbo-al-espacio-sistema-de-satelites-morelos.

Shorter, D. D., and K. TallBear. 2021. "An Introduction to Settler Science and the Ethics of Contact." *American Indian Culture and Research Journal,* 45 (1), 1–8.

Simpson, L. B. 1966. *Many Mexicos* (4th ed.). University of California Press.

Singh, V. 2023. "Spacefaring for Kinship." In *Reclaiming Space: Progressive and Multicultural Visions of Space Exploration,* edited by J. Schwartz, L. Billings, and E. Nesvold, 94–101. Oxford University Press.

Smiles, D. 2020. "The Settler Logics of (Outer) Space." *Society and Space,* October 26, 2020. https://www.societyandspace.org/articles/the-settler-logics-of-outer-space.

Solano Rojas, A. 2018. *Playgrounds del México moderno.* Promotora Cultural Cubo Blanco.

Solís, F. 2000. "La Piedra del Sol." *Arqueología mexicana* 41, 32–39.

Solís, F. 2001. "Hablemos del Posclásico Tardío." *Arqueología mexicana* 50, 20–29.

Spivak, G. C. 2003. *Death of a Discipline.* Columbia University Press.

Spry, J. 2024. "Navajo Nation Objects to Private Moon Mission Placing Human Remains on the Lunar Surface." *Space.com,* January 4, 2024. https://www.space.com /moon-navajo-nation-objection-human-remains-ula-vulcan-centaur-celestis -elysium-space.

Stengers, I. 2005. "The Cosmopolitical Proposal." In *Making Things Public: Atmospheres of Democracy,* edited by B. Latour and P. Weibel, 994–1004. MIT Press.

Stengers, I., B. Massumi, and E. Manning. 2009. "History Through the Middle: Between Macro and Mesopolitics—An Interview with Isabelle Stengers." *Inflexiones: A Journal of Research Creation* 3. https://www.inflexions.org/n3_stengershtml.html.

Stewart, K. 2010. "Afterword: Worlding Refrains." In *The Affect Theory Reader*, edited by M. Gregg and G. Seigworth, 339–53. Duke University Press.

Stewart, K. 2011. "Atmospheric Attunements." *Environment and Planning D: Society and Space*, 29 (3), 445–53.

Stewart, K. 2019. "The Life of the Milieu." *History of Anthropology Newsletter* 43. https://histanthro.org/notes/life-of-the-milieu/.

Stuart, D. 2021. *King and Cosmos: An Interpretation of the Aztec Calendar Stone.* Precolumbia Mesoweb Press.

Szerszynski, B. 2019. "Epilogue: Indigenous Worlds and Planetary Futures." In *Indigenous Perceptions of the End of the World: Creating a Cosmopolitics of Change*, edited by R. Bold, 203–10. Palgrave MacMillan.

Szolucha, A. 2023. "Planetary Ethnography in a 'SpaceX Village.'" In *The Routledge Handbook of Social Studies of Outer Space*, edited by J. F. Salazar and A. Gorman, 71–83. Routledge, Taylor and Francis Group.

Taylor, D. 2003. *The Archive and the Repertoire: Performing Cultural Memory in the Americas.* Duke University Press.

Thurner, M., and J. Cañizares-Esguerra. 2023. *The Invention of Humboldt: On the Geopolitics of Knowledge.* Routledge.

Tingley, B. 2024. "NASA Responds to Navajo Nation's Request to Delay Private Mission Placing Human Remains on the Moon." *Space.com*, January 4, 2024. https://www.space.com/nasa-responds-navajo-nation-objection-human-remains-moon.

Torres, A. 2006. "Rufino Tamayo." *Decires*, 8 (8), 9–21.

Torres, A. 2012. "Construcción de identidades visuales: Rufino Tamayo." *Revista de la Universidad Crisóbal Colón*, 28, 88–106.

Torres, A. 2021. "Montajes museográficos y quiebres curatoriales: Expo 67 Montreal." In *Diplocmacia cultural en México durante la Guerra Fría: Exhibiciones y prácticas artísticas, 1946–1968*, edited by D. Cruz Porchini, C. Garay Molina, and M. Velázquez Torres, 55–75. Secretaría de Relaciones Exteriores.

Trejo Barrientos, L. 2008. "Muerte, ritual y sexo entre los zoques de los Chimalapas." In *Los zoques de Oaxaca: Un viaje por los Chimalapas*, edited by L. Trejo Barrientos and M. Alonso Bolaños, 335–80. INAH.

Treviño, N. 2023. "Coloniality and the Cosmos." In *The Routledge Handbook of Social Studies of Outer Space*, edited by J. F. Salazar and A. Gorman, 226–37. Routledge, Taylor and Francis Group.

Tronchetti, F., and H. Liu. 2021. "Australia's Signing of the Artemis Accords: A Positive Development or a Controversial Choice?" *Australian Journal of International Affairs* 75 (3), 243–51.

Troncoso Pérez, R. 2011. "Nepantla, una aproximación al término." In *Tierras prometidas. De la colonia a la independencia*, edited by B. Castany Prado, B. Hernández, G. Serés Guillén, and M. Serna Arnáiz, 375–98. Universidad Autónoma de Barcelona.

Tsing, A. L. 2011. *Friction: An Ethnography of Global Connection.* Princeton University Press.

Tsing, A. L. 2015. *The Mushroom at the End of the World: On the Possibility of Life in Capitalist Ruins.* Princeton University Press.

Urrutia Fucugauchi, J. 1999. "El Año Geofísico Internacional 1957–1958 y los programas de investigación interdisciplinaria en el inicio del siglo XXI." *GEOS,* June 1999, 128–31.

Vallejo, P. 2024. "¿Qué ha sucedido on la misión 'Colmena' y sus micro robots?" *Ecoosfera,* January 15, 2024. https://ecoosfera.com/cosmos/que-ha-sucedido-con-mision-colmena/.

Vaughn, M. K. 1997. *Cultural Politics in Revolution: Teacher, Peasants and Schools in Mexico, 1930–1940.* University of Arizona Press.

Verhoeven, Paul, dir. 1990. *Total Recall.* Carolco Pictures.

Walls, L. D. 2009. *The Passage to Cosmos: Alexander von Humboldt and the Shaping of America.* University of Chicago Press.

White, F. 2014. *The Overview Effect: Space Exploration and Human Evolution.* American Institute of Astronautics and Aeronautics.

White, F. 2019. *Cosma Hypothesis: Implications of the Overview Effect.* Strauss Consultants.

Zabusky, S. E. 1995. *Launching Europe: An Ethnography of European Cooperation in Space Science.* Princeton University Press.

Zamora, L., and M. Kaup. 2010. "Baroque, New World Baroque, Neobaroque. Categories and Concepts." In *Baroque New Worlds: Representation, Transculturation, Counterconquest,* edited by L. Zamora and M. Kaup, 1–35. Duke University Press.

INDEX

References to illustrations appear in italics. References to the notes indicate page and note number, thus: 246n19 is note 19 on page 246. For abbreviations and acronyms, please refer to the beginning of the book.

aerospace industry, 5, 101–2, 121, 246n19. *See also* Blue Origin; rockets and rocketry; SpaceX

ahueques, 161

Aldana y Villalobos, Gerardo, 36, 38, 48

Alemán Velasco, Miguel, 136–37

aliens, 10, 16, 70, 127, 171, 173–75, 181, 182, 202, 247n3. *See also* science fiction; UFOs

Álvarez, Román, 73–74

analog missions, 114–20, *118*, 122, 149, 155–58, *156*, 243n18

ansibles, 188

Anzaldúa, Gloria, 11, 60

Apollo Moon landings: Indigenous interpretations of, 164–65; Mexican responses to, 22, 26–27, 70, 101, 104, 134, 136–39, 144–46, 245n14; training missions for, 115, 155, *156* (*See also* analog missions)

archaeoastronomy, 31–32, 33, 36–39

architecture: astronomical orientations of, 31–32, 33, 37, 38 (*See also* observatories; pyramids)

Armas, Marcela, 129, 130

Armenta, Víctor Joel, 198

Armstrong, Neil, 67, 104, 117, 144, 145, 245n10. *See also* Apollo Moon landings

Arreola, Mario, 67–68, 156

Artemis Accords (2021), 22, 134, 146–47. *See also* space law; treaties

Artemis mission, 133, *134*. *See also* Moon

artificial intelligence, 192, 193, 207

asteroids, 149, 155

Astorga Poblete, Daniel, 48

Astrobotic, 148, 151–54

astrology, 26, 47, 55–56, 170, 240n12

astronauts. *See also* Armstrong, Neil; space passengers; space tourists: analog astronauts, 117–18; artistic representations of, 65–66, *66*, 67–68, *69*, 126,

astronauts (*continued*)
    242n7; citizenship of, 21, 99–100, 104,
    241n1; definition of, 101–3; food eaten
    by, 12, 83, 105, 142, 202; Indigenous
    accounts of, 164–65; Mexican, 101–8,
    112–14, 116, 128 (*See also* Echazarreta,
    Katya; Hernández, José; Neri Vela,
    Rodolfo); Mexican-American, 242n4;
    Pakal, the Mayan "astronaut," 5–6, 133,
    134; requirements for, 21, 79–80, 99–
    100, 104, 241n1; as role models, 116–17,
    138; training of, 156, *156* (*See also*
    analog missions); women as, 241n1,
    241n3, 242n4
astronomy, 5–6, 31–32, 95, 170–71, 214–
    17, 228, 249n5
astrotourism, 20, 23–24, 221. *See also*
    dark skies; stargazing
Atlantis (space shuttle), 12, 78, 79–83, 102.
    See also space shuttles
atmospheric studies, 71, 72, 86
Austin, Alfredo López, 35
Autonomous InterGalactic Space Pro-
    gram, 65–66, *66*
AzTechSat satellites, 88. *See also* satellites
Aztecs. *See also* Mesoamerica; Mexicas;
    Tenochtitlan: as cultural heritage, 17,
    32, 39, 55, 144, 149, 179, 185, 191

Balam (Mayan deity), 33
balloons, atmospheric and stratospheric,
    6, 72, 148, 207
baroque ethos, 11, 60–61, 203–4
Barranca de Metztitlán, 224
beans, 199, 200, 202, 223
Becerril, Jesús, 245n14
Bess, Cameron and Lane, 107
Bezos, Jeff, 103, 112. *See also* Blue Origin
Blue Origin flights, 102, 106, 107, 108–9
Boltvinik, Ilana, 97–98, *98*
Bonampak, 33, 38
borders. *See also* nepantla: crossing, 235,
    250n14; effects on Indigenous peoples,

    246n29; invisibility from space, 105,
    110–11, 114; and the space community,
    21, 99–101, 104, 119, 127–28, 146,
    241n1
Borges, Jorge Luis, 133, 148
Bradbury, Ray, 26, 127, 186, 205
Branson, Richard, 102

Cabo Tuna, 71–72, 120
Cabrera, Gustavo, 155
calendar systems, 34–35, 42, 47–48, 57–
    59, 60, 179–80, 240n8
Canada, 6, 74, 147, 212, 243n24
Candiani, Tania, 129, 130–31, *131*
Canguilhem, Georges, 15
Cantinflas, 137
Cape Canaveral, 71, 78, 88, 151
caravels, 17, 67–68, *69*
Carpio, Mariel, 26–28, 31–34, 239n2
Carta de Cielo/Sky Map project, 216
"case of the forty-three," 131
caves, 50, 51, 161, 188, 229
Celestis, 154
cenotes, 50, 51, 52
Challenger (space shuttle), 83. *See also*
    space shuttles
Chao, Marcela, 157, 182–84, 220–21.
    See also Marsarchive.org; Martelolco;
    Martenochtitlan
charros, singing, 75, 173, 176
Chichén Itzá, 33, 38, 78
Chilam Balam, 48, 240n10
Chimalpahin, 57
China, 145–46, 245n16
Christianity, 51–52, 54–63, 165, 195,
    245n14
cinema, 173–77, 180–82, 202, 205, 240n5,
    247n3
citizen scientists, 91, 116, 119
citizenship requirements, 21, 99–100, 104,
    119, 241n1
Claudio, Juan, 184–85, 200
climate change, 8, 9, 14, 86, 163, 190

Coatlicue, 44–46, *46*, 47, 179, 191
codices, 36–37, 48, 49, 57–58, 60, 179, 186, *189*, 190, 195, 240n10
Colmena projects, 22–23, 148–55, *150*
colonization: critiques of, 3, 27, 182–84, 186, 190; of Mars (*See* Mars); as negotiated processes, 58–63; as theme in space exploration, 10–11, 27, 144–45, 182–84
Columbus, Christopher, 8, 68, 137, 190
Comarca Minera region, 221–26. *See also* Peña del Aire; San Sebastián
comets, 36, 49, 53–56, 78, 245n14
Commercial Lunar Payload Services (CLPS), 151, 154
Committee for the Peaceful Uses of Outer Space (COPUOS), 122, 140, 147
conquest (of space), 10–11, 176–77, 182–84, 202–3
constellations, 26, 32, 33, 36, 38, 54, 65, *66*, 229, 234. *See also* stargazing; satellite constellations (*See* satellites)
Correa, Víctor, 108–9
Cortés, Hernán, 54, 240n11
cosmologies, 21, 49–52, 56, 158–63, 164–65
cosmonauts, 70, 79, 125, 129–31, *131*, 184. *See also* astronauts
Council of Science and Technology (CITNOVA), 207–8, 221, 228–29
Cuauhtémoc (emperor), 143, 194
CubeSats, 86–89, 102. *See also* satellites
Cuevas Méndez, Juan Carlos, 133–34, *134*
Curiosity (Mars rover). *See* robots and rovers
Cydonia Foundation, 119

Dalton, David S., 176–77
darkness, 208–10, 213, 232–33, 235, 249n7. *See also* night
dark skies, 19, 27, 78, 207, 217–18, 219–20, 227–36
DarkSky International, 219, 227, 235

decoloniality, 11, 18, 119, 128, 145, 184–85, 190, 203
Dereum Labs, 120–21
deserts, 7, 10, 155–58, *156*, 169–71, 246n29
Díaz, Porfirio, 40, 216, 239n7, 246n18
Díaz Álvarez, Ana Guadalupe, 50, 51
Díaz Cruz, Rodrigo, 61
Díaz de Castillo, Bernal, 194
Díaz Infante, Juan José, 91–92, 129, 130–31, *131*
Díaz Ordaz, Gustavo, 75–76, 136, 138, 139, 197
Discovery (space shuttle), 104. *See also* space shuttles
Dovey, Ceridwen, 112
Duarte, Carlos, 147
duendes (gnomes), 220, 221, 222, 232
Durán, Diego, 59
dust, 16, 149, 150, 152, 186, 188

eagles, 28, 41, 159, 186, 203, 229
Earth: prioritizing, 27–28, 158; viewed from space, 14, 77, 88–89, 94, 107, 109–14, 117, 143 (*See also* satellites); views of space from, 114, 158 (*See also* dark skies; observatories)
earthquakes, 80–81, 98, *98*, 116, 198
Echazarreta, Katya, 102, 106–8, 113, 114, 128, 218
Echazarreta Foundation, 113, 114
de Echeverría, Bolivar, 11, 177, 203
Echeverría García, Jaime, 162
eclipses, 33, 36, 37, 44, *134*, 142, 162–63, 169–71, *171*
ecotourism, 23–24, 224–26, 235, 250n14
ejidatarios, 205, 222–23, 227, 229–31
ejidos, 13, 208, 222–23, 226, 229–31
electrification, 210–14
English, fluency in, 5, 79, 100, 104, 119, 149, 151, 244n1
environmental monitoring, 14, 77, 88–89, 94. *See also* satellites

ethics, place in space exploration, 19, 27, 125, 126, 154–55, 202–3
Eutelsat, 85, 86, 87, 92–93
Explorer 1, 71. *See also* satellites

failure, 151–54, 157
Félix, Carmen, 99–100, 114, 116, 241n1
fieldwork, 18–20
folk dancers, 75, 82–83, 145
France, 11, 100, 121, 215, 242n3
frontera, 127. *See also* borders; frontier
frontier, space as, 10, 12, 127, 149–50
Fuentes, Carlos, 144–45

Gagarin, Yuri, 70, 125, 184
Galindo, Jesús, 37
Galinier, Jacques, 160
Gall, Ruth, 73–74
Gallo, Rubén, 196
Genauer, Emily, 142
geological heritage, 224–25, 249n10, 250n11. *See also* meteors and meteorites; mining
glyphs, 33, 36, *189*, 190
Grajeda, Genaro, 116–17, *118*
gravity, 129–31, *131*
Green Revolution, 223–24
ground tracking stations, 70, 73, 76, 80, 90, 94–96, *95*, 98, *98*, 102, 241n8
Gruzinski, Serge, 181
Guzmán Huerta, Rodolfo ("Santo"), 175–77. *See also* lucha libre

habitats, 119, 193, 196
HemisFair (1968), 140–41
Hernández, José, 102, 104–6, 113, 146, 242n7
Hernández Márquez, Juanita, 241n3
Hñahñu (Otomí), 160–63, 226
Huasca de Ocampo, 224–26, 230
Huicholes, 235
Huitzilopochtli (Aztec war deity), 17, 41–43, 44, 191, 248n13

Humberto, Rey, 27
Humboldt, Alexander von, 8, 39, 44, 45, 46, 190, 220, 222, 224, 240n7

imaginaries, 17–18, 20, 22, 31, 68, 101, 127, 135, 157, 166. *See also* Martelolco; Martenochtitlan; utopias; worlding
immigration, 223, 235, 236, 250n14
Imperial College of the Holy Cross of Tlatelolco, 194–95
Indigenous peoples: approaches to science, 11, 163; as foundation for imagined futures, 28, 29–30, *30*, 190–92; perspectives on space exploration, 154, 164–65, 202–3
Intelsat satellites, 74–75, 76. *See also* satellites
International Astronautical Congress (IAC), 85, 100, 115–16, 180
International Astronomical Union (IAU), 179, 244n1, 247n5. *See also* naming
International Geophysical Year (IGY), 4, 70–71
International Space Station, 13, 68, *69*, 88, 97, 102, 103, 104–5
International Traffic in Arms Regulations (ITAR), 115, 122
"Interplanetary Simulations," 156–58. *See also* Terremoto

jaguars, 17, *30*, 33, 42, 121, 159

Kármán line, 72, 102–3
Kino, Eusebio, 56, 246n29
Kirchoff, Paul, 34
KOSMICA Institute, 129–31, *131*, 169–71, *171*

Laboratory of Space Instrumentation (LINX), 22–23, 89–92, 148–55, *150*. *See also* National Autonomous University of Mexico (UNAM); National Laboratory for Access to Outer Space (LANAE)

ladders, 67–68, *69*, 134, *135*
Landsat satellites, 77. *See also* satellites
La Raza metro station, 26, 28
launch sites, 11, 71–72, 78, 88, 120, 146, 151, 207, 242n3
León-Portilla, Miguel, 38, 49, 59–60
Levi, Heather, 175
Ley del Cielo/Law of the Sky, 217, 218
lighting, artificial, 208–14, 219–20, 228, 233, 234, 235
light pollution, 16, 23, 32, 78, 211, 213–14, *214*, 216, 217–20, 228
lights in the sky, 6, 26, 221, 232–33, 235. *See also* satellites; UFOs
Lorente, David, 159, 161
lucha libre, 173–77, 191, 192, 247n2
Lugo, Helena, 157–58
Luis Enrique Erro Planetarium, 26, 31–32, 70. *See also* planetariums

Maffie, James, 47, 52–53, 60
maize, 34, 37, 65–66, *66*, 192
malinalli, 47
manifest destiny, 10–11, 182. *See also* conquest (of space); frontier, space as
maps and mapping. *See also* naming: community maps, 229–30; cosmographies, 12, 50–52, 56, 57, 123, 159–60; imaginary maps, *178*, 178–79, *189*; of outer space, 12; of the physical sky, 95, 216; use of astronomy in, 215
maquilización, 101, 115, 153, 182
Marcus, George, 19
Mars: analog missions to, 116–17, *118*, 119–20, 156, 243n18; deities and symbology associated with, 33; future imaginaries of, 3, 17, 85, 114–20, *118*, 180–84, 188, 198–99 (*See also* Marsarchive.org; Mars in Guerrero; Martelolco; Martenochtitlan); geographies of, 178–79; Mexican commentaries on colonization of, 3, 23, 178–80, *180*, 190–91, 192–93, 203–4; observation

of, 37; in popular culture, 173–77, 180–82
Marsarchive.org, 17–18, 19, 23, 170–80, *178*, 183–84, 203–4, 220. *See also* Martelolco; Martenochtitlan
Mars Desert Research Station (MDRS), 116, 117, 118, 119
Marshall, Albert D., 163
Mars in Guerrero workshop, 23, 202–3. *See also* Marsarchive.org; Martelolco; Martenochtitlan
Mars Station/Estación Marte, 205–6, *206*
Martelolco, 23, *178*, 178–80, 193–94, 198–201. *See also* Marsarchive.org; Mars in Guerrero workshop; Martenochtitlan; Tlatelolco
Martenochtitlan, 3, 23, *178*, 178–80, 184–93, *189*. *See also* Marsarchive.org; Mars in Guerrero workshop; Martelolco; Tenochtitlan
Martians, 173–77, 181, 182. *See also* aliens; Mars in Guerrero workshop; Martelolco; Martenochtitlan
Masewal (Nahuas), 163
Matters of Gravity collaboration, 129–31, *131*, 244n1
Mayan Archaeoastronomy (educational film), 32–33, 48
Mayapán, 37
Mayas, 5–6, 32–34, 47–48, 133, *134*, 247n32
Mayrán, 169
Medina Tanco, Gustavo, 89, 91, 148–55, 166
Mejuto, Javier, 162
Mendieta, Francisco Javier, 79
Mennonites, 6–7
Mercury 13, 242n13. *See also* Project PoSSUM/PoSSUM 13
Mesoamerica: archaeoastronomy of, 31–32, 33, 36–39; conceptualizing and defining, 34–35, 239n6; cultural references to, 17–18, 78–79, 177–80,

Mesoamerica (*continued*)
178, 186–87, *189*, 190–93, 226, 240n5
(*See also* Martelolco; Martenochtitlan;
Tenochtitlan; Teotihuacán); time-place
concepts of, 47–52; vertical cosmos of,
35, 49–52
Messeri, Lisa, 12, 119, 179
mestizaje, 17, 28–31, *30*, 187, 193, 202–3
meteors and meteorites, 15, 36, 161, 169,
191, 228, 241n4, 249n10
MEX-1 analog mission team, 116–17, *118*,
120–22. *See also* analog missions
Mexican Aerospace Fair (FAMEX), 101–2,
121
Mexican Encounter of Experimental
Rocket Engineering (ENMICE): rocketry competition, 4–7, *7*, 125
Mexican flag, worn or displayed by astronauts, 67, 68, *69*, 104–5, 106
Mexican Institute of Geography and
Statistics, 215
Mexican Revolution (1910–1920), 5, 28–
29, 195, 222, 248n15
Mexican Space Agency (AEM): and art
and artists, 67–68, *69*, 133–34, *134*,
146, 183; employment opportunities
with, 100, 116; institutional history of,
74, 85, 96, 128, 147; public outreach of,
88, 124, 133–34, *134*, 146
Mexican Space Collective (CEM), 92
Mexican Space Program (PEM), 96
Mexicas, 47, 49. *See also* Aztecs
"Mexico, I Will Come on Board Your
Ship!," 67–68, *69*
Mexico City. *See also* Tenochtitlan;
Tlatelolco: architecture of, 180–82
(*See also* Nonoalco Tlatelolco Urban
Complex; Plaza of the Three Cultures);
earthquake damage to, 80–81, 98, *98*,
198; public lighting of, 209–14; urban
development of, 39, 44, 248n3
MEXSAT satellite system, 84, 96, 128. *See
also* satellites

milieux, 3–4, 7–9, 15–18, 101, 133–35
Milky Way, 36, 38, 45, 234
mining, 211, 215, 218, 221–22, 224
missionaries, 51–52, 56–57
Moctezuma II, 54, 144
modernity: critiques and engagement
with, 29–30, 70, 73, 141, 203–4, 212;
place of technology in, 74–75, 78, 210–
13; visual representations of, 74, 75,
176, 180–82
Monsiváis, Carlos, 81, 145, 176
de Montúfar, Alonso, 40–41
Monument to La Raza, 28
Moon: analog missions to, 115, 118, 149;
animals associated with, 17, *30*, 33, 42,
121, 133, *134*, 135, 141, 159, 247n32;
artistic representations of, 65, *66*,
133–34, *134*, 142, 245n14; colonization
of, 144–45; competing understand-
ings of, 23, 135, 154, 155, 165–67, 171;
cosmological location of, 49, 164–65;
deities and symbology associated
with, 33, 45, 165, 247n32; and eclipses
(*See* eclipses); as focus of imaginaries,
22–23, 133–35, *134*; human activity
on, 22–23, 26–27, 70, 102, 112, 120–
21, 133, *134*, 135, 139–40, 148–55,
150; legal status of, 166; mythological
origins of, 43, 158–59, 247n32; observation of, 36, 37, 38; paired with the
Sun, 158–59, 160, 161, 162; personal
relationships with, 135, 166, 171; phenomena associated with, 159–62; place
in Indigenous cosmologies, 158–63; as
source of light at night, 209, 234
Moon Treaty, 140, 146. *See also* space law;
treaties
Morelos satellites, 21–22, 77–79, 80, 81,
84–85, 94, 98, *98*. *See also* satellites
murals: depictions of astronauts, 102, 108,
141; depictions of local landscapes,
221; depictions of space exploration,
70, 245n14; depictions of telescopes,

228; invocation of *la raza cósmica*,
28–31, 141–44; Mexican School, 28–
31, *30*, 141; as social critique, 29–31,
142–43
Musk, Elon. *See also* SpaceX; Starlink:
Mars colony ambitions of, 3, 17, 85,
115, 180, 182–84, 188, 191–93, 204;
Tesla factories of, 115, 243n15

Nahuas, 47, 49, 52–53, 60–61, 202–3
Nahuatl, 17, 149
Nahum, 129, 130, 131, *131*, 170. *See also*
KOSMICA Institute
naming. *See also* place-making: of devices
and missions, 72, 88, 149, 150, 187; of
objects and places in space, 10, 17, 37,
179, 203, 244n1, 247n5
NanoConnect-2 satellite, 89–91. *See also*
satellites
La NASA no es la raza installation, 97–
98, *98*
National Aeronautics and Space Administration (NASA). *See also* Apollo
Moon landings; astronauts; space
shuttles: and Artemis Accords, 22,
134, 146–47; citizenship requirements
of, 21, 99–100, 241n1; collaborations
with, 5, 71, 79–83, 142; Commercial
Lunar Payload Services (CLPS), 151,
154; internships at, 100, 106, 122, 127;
Landsat system of, 77; Mars rovers,
183, 184, 205
National Astronomical Observatory
(OAN), 123, 216, 217, *219*. *See also*
observatories
National Autonomous University of
Mexico (UNAM), 20, 26, 71, 75–76,
88, 148, 221, 224, 228–29. *See also*
NanoConnect-2 satellite; National
Laboratory for Access to Outer Space
(LANAE); Project Colmena
National Commission for Outer Space
(CONEE), 72, 73–74

National Institute of Astronomy, Optics,
and Electronics (INAOE), 123, 125,
137, 217, 249n5
National Laboratory for Access to Outer
Space (LANAE), 148, 207
National Polytechnical Institute (IPN), 26,
75–76
Navajo Nation, 154
Navarrete, Federico, 25, 44, 45
nepantla, 11, 21, 47, 59–62, 101
Neri Vela, Rodolfo, 12, 78–83, 102, 104,
105–6, 116, 127, 128, 148. *See also*
astronauts
NewSpace, 3, 23, 85–89, 101, 120–22, 203
New Spain: as existing in a state of nepantla, 59–62; place of portents and
astrology in, 55–56
Nezahualpilli, 54
night and the night sky, 26, 27, 33, 43–44,
50, 233, 236–37. *See also* dark skies
night hikes, 231–36, 250n14. *See also*
astrotourism; stargazing
Noche de las Estrellas/Night of the Stars
festival, 228
nocturnal activities, 208–10, 213, 231–36.
See also night hikes; stargazing
Nonoalco Tlatelolco Urban Complex,
195–96, 197, 201
nopal cactus, 107, 187, 191, 192, 193, 234
nuclear weapons, 139, 174, 176

observatories, 123, 215–16, 217–18, *219*,
249n5. *See also* archaeoastronomy;
astronomy; pyramids
Ojeda, Oscar, 118–19
ollin, 47
Olson, Valerie, 13, 193
Olvera, Ana Cristina, 157–58
Olympic Games (1968), 75–76, 140–41.
See also Tlatelolco massacre
Opportunity (rover), 205. *See also* robots
Orion (constellation), 33, 38
Orozco, José Clemente, 29

Our World (1967), 74–75, 176, 197
outer space: boundaries of, 102–3; com-
    mercialization of, 112, 124, 147–48,
    149–50, 151, 155; military involvement
    in, 71, 87, 112, 243n20, 243n27; own-
    ership of, 124; regulation of, 147–48,
    155 (*See also* space law)
Outer Space Treaty (1967), 22, 139–40
overview effect, 107, 109–14, 143

Padilla, Andrés, 206
Padilla, Rosa Inés, 26–28, 31–34, 135, 148,
    239n2
Painani-1, 87. *See also* satellites
Pakal (the Mayan "astronaut"), 5–6, 133,
    134
Pani, Mario, 197. *See also* Nonoalco
    Tlatelolco Urban Complex
Paz, Octavio, 46, 145, 246n28
Peña del Aire, *225*, 226–31, 250n14. *See
    also* Comarca Minera region; San
    Sebastián
Peralta y Fabi, Ricardo, 79
Picazo, David, 223–24
Piedra del Sol, 39–44, *40*
Piña, Yair, 117–18
Pinacate Biosphere Reserve, 155–58
piñatas, 130–31, *131*, 180
place-making, 12–14. *See also* naming;
    worlding
planetariums, 25, 26–28, 31–34, 65–66,
    66, 108, 213–14, 218
planetary (concept), 13–14
Plaza of the Three Cultures (Tlatelolco),
    194, 196–97, 200–201
Pleiades, 33, 36, 38
portents, 53–56
Povinelli, Elizabeth, 224
Project Colmena, 22–23, 148–55, *150*
Project PoSSUM/PoSSUM 13, 114,
    242n13. *See also* analog missions
prostheses, satellites and robots as, 16,
    68, 91

de la Puente, Ale, 129, 130–31, *131*
Puig, Iván, 206
Purépechas (Tarascos), 162–63
pyramids, 17, 27, 28, 31–32, 50, 52. *See
    also* archaeoastronomy

Quetzalcóatl, 32, 37, 41–43, 54, 145, 187,
    193, 200, 234, 247n32
QuetzSat satellite system, 83–84. *See also*
    satellites
quincunx, 41

rabbits, 33, 133, *134*, 135, 141, 159,
    247n32
radio broadcasts, 136, 138–39
Rarámuris (Tarahumaras), 161–63, 235,
    247n32
la raza cósmica, 21, 28–31, 239n3
reciprocity, 16, 44, 60, 163, 166
Conde de Regla, 222
Count of Revillagigedo, 39, 209
Revolutionary Institutional Party (PRI),
    79–80
Rivera, Diego, 29, 200, 240n4
robots and rovers: artistic rovers, 205,
    206; lunar robots, 22–23, 120–21,
    148–55, *150*; Mars rovers, 183, 184,
    199, 205; naming, 17, 149, 150; as
    prostheses, 16
rockets and rocketry. *See also* Blue Origin;
    SpaceX: artistic representations of,
    65–66, *66*, 67, 126, 133, 141, 144, 173,
    196, 197, 245n14; explosions and de-
    bris, 83, 94; Mexican interest in, 4–7,
    7, 70, 71–73; naming of (*See* naming);
    rocketry clubs and competitions, 4–7,
    7, 125
Romero, Laura, 233
Romero de Terreros, Pedro. *See* Regla
Ross, Amadís, 184–85
rovers. *See* robots and rovers
Russia, 93, 129–31, *131*, 241n4. *See also*
    Soviet Union

sacrifice, ritual, 44, 52
Saldaña, Vicente, 76
San Sebastián, 208, 222–23, 229–31. *See also* Comarca Minera region; Peña del Aire
Santo el Enmascarado de Plata, 173–77. *See also* lucha libre
Santo vs the Martin Invasion (1967), 173–77
sargasso, 7–9, *15*, 16
satellites. *See also* Sputnik; telecommunications: artistic satellites, 91–92, 94; citizen satellites, 92; constellation systems, 86, 88; kit satellites, 241n4; microsatellites, 86–89; Mexican control of, 77–79, 83–84, 88, 89–90, 102, 241n4 (*See also* CubeSats; MEXSAT satellite system; Morelos satellites; NanoConnect-2 satellite; Solidaridad satellite system); artistic representations of, 17, 65, *66*, 67–68, *69*, 126, 141, 176; changes in technology, 83–85, 86–88; displaying, 70, 102; environmental monitoring with, 7–9, 16, 77, 88–89, 94; foreign ownership of, 74–75, 76, 77, 83–84, 93, 112, 158, 243n15; launching, 71, 80, 81, 88, 89–90, 92, 146, 241n4; military uses of, 71, 87, 243n27; naming of, 17, 88 (*See also* naming); nobility of, 84, 87; personal connections to, 89–92; perspectives from, 68, 74, 75; poetic potential of, 90–92; positions of, 77, 93, 103; as potential space junk, 86, 92–94; as prostheses, 16, 68, 91; regulation of, 93–94, 112, 147; technological sovereignty and, 21–22, 68, 77, 80, 81, 83–84, 88–90, 100, 146; (in)visibility of, 23, 32, 92–94, 158
SATMEX satellite system, 83–84. *See also* satellites
SatNOGS network, 90. *See also* ground tracking stations; satellites

Schmelz, Itala, 177, 192
science fiction, 16, 19, 20, 27, 104, 127, 180–82, 185, 186, 188, 202, 244n2
Secretary of Communication and Transportation (SCT), 76–80
Secretary of Tourism (SECTUR), 231, 232
SEFT-1 project, 206
serpents, 17, 28, 36, 37, 38, 41, 43, 45, 54, *178*, 187, *189*, 200, 248n9. *See also* Quetzalcóatl
Shaw, Brewster, 81–82
Sheinbaum, Claudia, 96, 201
Siempre!, 144–45
de Sigüenza y Góngora, Carlos, 56
Singh, Vandana, 111
Siqueiros, David Alfaro, 29, 143–44, 240n4
Skylab, 68, *69*. *See also* space stations
Sky Quality Meters (SQMs), 227–28, 236
social media, 106, 108, 128, 153
Solidaridad satellite system, 67–68, *69*, 83, 84, 94. *See also* satellites
Solís Wolfowitz, Vivianne, 79
Soviet Union, 29, 70, 71, 72, 139, 144, 196, 245n17. *See also* Russia
space camps, 108, 114, 128
space community: borders and divisions within, 22, 111–13, 119–22, 125, 126–28; women's place in, 113, 114, 120, 124, 241n3, 242n13; young people's place in, 22, 99–101, 113, 116, 120–22 (*See also* Space Generation; Space Generation Advisory Council)
space exploration. *See also* analog missions: artistic depictions of, 65–66, *66*, 67–68, *69*, 70; commercialization of, 112, 120–21, 124, 155, 201; conquest as theme in, 10–11, 27, 176–77, 182–84, 202–3; English fluency in, 5, 79, 100, 104, 119, 149, 151, 244n1; ethics of, 19, 27–28, 111, 125, 126, 154–55, 202–3; national budgets for, 85, 115, 116, 153–54

Space for Humanity (S4H), 107, 109–10, 112

Space Generation, 20, 99–101, 116–17, 120–22, 126–28. *See also* Space Generation Advisory Council (SGAC)

Space Generation Advisory Council (SGAC), 19, 22, 100, 119, 122–26, 183–84, 243n24, 243n25. *See also* Space Generation

space junk, 86, 92–94, 97–98, *98*, 152, 241n7

space law, 72, 93–94, 147–48, 155. *See also* treaties

space passengers, 102, 106, 107, 108–9. *See also* astronauts; space tourists

Space Race, 21–22, 71–73

spaceships, 5–6, 65–67, *66*, 133, 141, 173, 175, 186–87, 190–93. *See also* rockets and rocketry; space shuttles

space shuttles: Atlantis, 12, 78, 79–83, 102; Challenger, 83; Discovery, 104; virtual model of, 126

space stations, 13, 68, *69*, 88, 97, 102, 103, 104–5

space tourists, 102, 103, 106, 107, 108–9. *See also* astronauts; astrotourism; space passengers

SpaceX, 11, 88, 94, 112, 115, 183, 187, 188, 192–93, 204

Spanish, cosmologies of, 51–52, 56–59

spatiality, Mesoamerican, 48–49

speculation, 11, 18–20, 23, 125, 199, 204. *See also* Martelolco; Martenochtitlan; science fiction; utopias; worlding

Spring, Sherwood "Woody," 82–83

Sputnik, 21, 70–71, 126. *See also* satellites

stargazing, 218–19, 228–29. *See also* dark skies; night and the night sky; night hikes

star glyphs, 33, 36

Starlink, 93, 112, 158, 243n15. *See also* Musk, Elon; satellites

stars. *See* constellations; Milky Way; night and the night sky; stargazing; Sun

Star Trek, 104, 117, 244, 244n2

Stengers, Isabelle, 17, 135, 166

Stewart, Kathleen, 12–13, 15, 16, 17

Sun: artistic representations of, 65, *66*; associated with eagles, 159; cosmological location of, 49; multiplicity of, 41–43; mythological origins of, 158–59, 247n32; observation of, 36; paired with the Moon, 158–59, 160, 161, 162; phenomena associated with, 160, 161

Sunstone. *See* Piedra del Sol

surveillance technologies, 6, 87, 209, 213, 216, 220, 249n7

synchrotron particle accelerator, 207–8

Szerszynski, Bronislaw, 14, 163

Tamayo, Rufino, 29, 30, *30*, 141–43

technological sovereignty, 21–22, 68, 77–78, 85–89, 96, 116–20, 124

technology. *See also* robots and rovers; rockets and rocketry; satellites; spaceships: artistic engagement with, 207–8, 212; critiques of, 175, 176, 201; Indigenous perspectives on, 154, 163

Telecomm, 77

telecommunications, 21–22, 72, 74–75, 77–81, 96, 100. *See also* satellites; television

telescopes, 26, 94–96, *95*, 219, 221, 228–29, 249n5

Telesistema Mexicano, 136–37. *See also* Televisa

Televisa, 77, 80–81, 98, *98*. *See also* Telesistema Mexicano

television, 71, 73–75, 136. *See also* telecommunications

Tenochtitlan, 28, 31–32, 39–40, 57, 208–9, 248n13. *See also* Martenochtitlan

Teotihuacán, 32, 38, 222

teotl, 47, 52–53, 56

terranauts, 110–11

Terremoto, 156–57

tesifteros, 161
time, 47–48, 57–58, 179–80. *See also* calendar systems
Tlahuizcalpantecuhtli (deity), 37
Tlaloc (rain deity), 37, 159
tlamatiliztli, 52–53
Tlatelolco, 194–98, 200–201, 248n13. *See also* Martelolco
Tlatelolco massacre, 75–76, 139, 197, 198, 200, 201, 248n15
Tlatelolco Urban Orchard, 198, 199, 200
Tohono O'odham (Papagos), 155, 246n29
Tonantzintla, 123–24, 216–17
Torquemada, Juan de, 54–55, 208–9
tortillas, 6, 12, 83, 105
Total Recall (1990), 180–82
tourism. *See* astrotourism; ecotourism; space tourists
Tower of Tlatelolco, 196, 198
transhabitation, 23, 193, 204, 248n12
Trapped, 97
treaties, space, 22, 70, 139–40, 146
Treaty of Tlatelolco, 139, 174, 176, 196–97
Trejo, Leopoldo, 164–65
TRES art collective, 84–85, 97–98, *98*
Troncoso Pérez, Ramón, 60
Trump, Donald, 117, 243n15
Tsing, Anna, 12, 14, 231
Tulancingo, 94–96, *95*, 241n8
Tunnel of Science, 26

UFOs, 10, 27, 199, 205, 221, 232
Ulises I satellite, 91–92, 94. *See also* satellites
UNAMSAT satellites, 241n4. *See also* satellites
UNESCO, environmental designations of, 155, 216, 217, 224
United States, Mexico's complicated relationship with, 5, 10, 17, 70, 73, 123, 145–46, 177, 181–82, 223, 239n7, 245n17. *See also* National Aeronautics and Space Administration (NASA)

Universum (science museum), 26, 32–33. *See also* planetariums
Utopía Libertad, 218–19
utopias, 29–30, 157, 180–82, 202–3, 218–19
Uxmal, 33

Valadés, Diego de, 56
Valdés, José, 88
Valverde, Miguel, 208
Vasconcelos, José, 28, 30, 141, 239n3
Venus, 33, 36, 37, 48, 49, 158, 166, 215, 247n5
Vielma, Carlos, 205–6
Villa, Pancho, 5
Viñas, Rodrigo, 97–98, *98*
Virgin Mary, 55, 61, 245n14
virtual reality environments, 102, 126, 180 184
Voyage to Mars exhibition, 177–80, *178*. *See also* Marsarchive.org

White, Frank, 109, 111
witches, as fireballs at night, 221, 232–33, 235
women, 113, 114, 120, 124, 213, 221, 231–32, 241n3, 242n13
worlding and reworlding, 12–14, 22, 114, 119–20, 128

Xólotl, 37

Zabludovsky, Jacobo, 137
Zapatista Army of National Liberation (EZLN), 65–66, *66*
Zapotecs, 48, 149
zero gravity, 129–31, *131*
Zeus rocket, 71–72. *See also* rockets and rocketry
Zoque, 164–65

ABOUT THE AUTHOR

Anne W. Johnson is a professor in the Graduate Program in Social Anthropology at the Universidad Iberoamericana in Mexico City. She has a
BA in anthropology and theater arts from Brown University, and an MA
and PhD in social anthropology from the University of Texas at Austin.
Her research interests include the anthropology of outer space, the social
studies of science, technology, and art, material culture, and performance
studies.